MW00679688

SELECTIVE COMPUTATION

Selective Computation

Richard E. Bellman

Professor of Mathematics
Electrical Engineering and Medicine
University of Southern California
Los Angeles, California 90007

Published by

World Scientific Publishing Co. Pte. Ltd.
P. O. Box 128, Farrer Road, Singapore 9128
242, Cherry Street, Philadelphia PA 19106-1906, USA

Library of Congress Cataloging in Publication Data

Bellman, Richard Ernest, 1920-1984
Selective computation.

Includes bibliographies.
1. Mathematical analysis. I. Title.
QA300.B453 1985 515 85-10696
ISBN 9971-966-86-7

Copyright © 1985 by World Scientific Publishing Co Pte Ltd.

All rights reserved. This book, or parts thereof, may not be reproduced in any form or by any means electronic or mechanical, including photocopying, recording or any information storage and retrieval system now known or to be invented, without written permission from the Publisher.

Printed in Singapore by General Printing and Publishing Services Pte. Ltd.

To Nina

PREFACE

This book is devoted to the following questions: How does one calculate what is desired without calculating a lot of data which is not? It is not to be expected that any general method will exist. Rather, one must use particular features of the problem under consideration. We have, however, employed a number of general techniques.

In Chapters 5, 7, 12, 13, and 14, we have employed semi-groups. We have employed the theories of dynamic programming and invariant imbedding throughout. Particular applications will be found in Chapters 3, 8, 9, 10, 11, and 13.

We have also employed the method of upper and lower bounds, particularly in Chapter 3. Throughout the book we point out whenever we can obtain an upper or lower bound.

We show that invariant imbedding and dynamic programming lead to partial differential equations in the consideration of some problems. Conversely, we can find particular values of the solutions of these partial differential equations by solving the problems by other methods.

We discuss the use of the Lagrange multiplier and of the maximum transform.

Let us now discuss the content of the book in more detail.

Chapter 1, is devoted to finding a particular element in the inverse of a matrix. To this end, we employ a formula from the theory of quasilinearization. We also give some other results connected with linear algebraic equations. In Chapter 2, we turn to the problem of finding the positive definite square root of a positive definite matrix. We employ a formula from control theory.

In Chapter 3, we give various methods for finding upper and lower bounds for the largest characteristic root of a positive definite matrix and the smallest characteristic root of a Sturm-Liouville problem. We also show that the largest characteristic root of a nonnegative matrix has a variational characterization which allows us to obtain upper and lower bounds. Numerical results are given in connection with some of these problems. In Chapter 4, we present another method for determining the smallest characteristic root of a Sturm-Liouville equation.

In Chapter 5, we turn to linear differential equations with constant coefficients, studying the matrix exponential. We give two methods for calculating this quantity and show how useful the powers of two are. Using the matrix exponential, we can obtain a formula for the inverse of a matrix which is occasionally useful. In Chapter 6, we give a number of results where the coefficients are time dependent. In Chapter 7, we study nonlinear differential equations. We show that the semigroup concept leads to questions of iteration. We also show that relative invariants can be useful in this context.

In Chapter 8, we study the problem of asymptotic behavior. We give a result using the extrapolation formula/Kronecker which is useful in many ways. Then we give two applications of invariant imbedding.

In Chapter 9, we show that dynamic programming applied to variational problems leads to important classes of partial differential equations. This means that we possess upper bounds for the solution of the associated partial differential equations using trial functions. In Chapter 10, we show that the use of invariant imbedding leads to important classes of partial differential equations. Thus, following the dictum of Jacobi, we can obtain particular values of the solution of the partial differential equations by attacking the original problem.

In Chapter 11, we use invariant imbedding to study the problem of maximum range. We point out that the same method may be used to study the problem of maximum penetration. In Chapter 12, we discuss invariant imbedding from a general point of view, semigroups in space, time and structure. In Chapter 13, we show that analogous functional equations can be obtained for variational problems. Various generalizations are indicated.

In the concluding chapter, Chapter 14, we study allocation processes by two techniques. First, we show that the use of Lagrange multipliers yields a great reduction in dimensionality, and then we discuss the use of the maximum transform to obtain an explicit solution.

It will be clear from the text how many open questions remain.

It is a pleasure to thank Rebecca Karush who prepared this manuscript.

Santa Monica, May 1982 Richard Bellman

CONTENTS

CHAPTER III: LARGEST AND SMALLEST CHARACTERISTIC ROOTS

CHAPTER IV: ON THE DETERMINATION OF CHARACTERISTIC VALUES FOR A CLASS OF STURM-LIOUVILLE PROBLEMS

CHAPTER V: LINEAR DIFFERENTIAL EQUATIONS WITH CONSTANT COEFFICIENTS

CHAPTER VI: LINEAR DIFFERENTIAL EQUATIONS WITH VARIABLE COEFFICIENTS

CHAPTER VII: NONLINEAR DIFFERENTIAL EQUATIONS

CHAPTER VIII: ASYMPTOTIC BEHAVIOR

CHAPTER IX: PARTIAL DIFFERENTIAL EQUATIONS AND DYNAMIC PROGRAMMING

CHAPTER XIV: ALLOCATION PROCESSES, LAGRANGE MULTIPLIERS AND THE MAXIMUM TRANSFORM

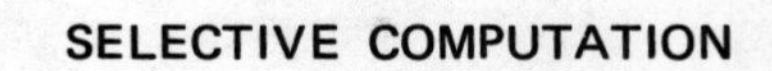
SELECTIVE COMPUTATION

Chapter I

LINEAR ALGEBRAIC EQUATIONS

1.1 Introduction

In this chapter, we want to consider some questions associated with linear equations.

First, we shall say a few words about Cramer's solution. Then we shall give the method of Gaussian elimination, pointing out some of the difficulties. Then we shall show how both procedures can be combined to evaluate determinants.

Following this, we turn to the principle objective of the chapter. We shall use a formula for the inverse of a matrix to obtain a particular element. We shall present the method first for symmetric matrices and then show how the result may be used to obtain a particular component of the solution of linear systems.

We point out that the method may be used for more general operators, and in particular, to study the Green's function.

Finally, we show how imbedding techniques may be used to handle complex operators.

1.2 Cramer's Solution

Consider the system of linear equations

(1.2.1) $Ax = b.$

Here, we suppose that the matrix A is nonsingular. Then, the solution may be written very elegantly as

$$x = A^{-1} b \tag{1.2.2}$$

The elements of A^{-1} may be determined by Cramer's rule in terms of determinants. This representation is very important theoretically, but not very useful computationally when the dimension of A is large.

As we know, a determinant of dimension n requires $n!$ operations if the usual method is employed. Each operation requires n multiplications. The number $n!$ commands a great deal of computational respect. We have

$$10! = 3,628,800 \quad .$$

A simple majorization shows that $20! = 10^{10} \times 10!$, a very large number indeed. Another useful measure is that a year has approximately $\pi \times 10^7$ seconds. Even at micro-second, or nano-second, speed, we cannot perform that many operations at present.

Furthermore, if we could perform that many multiplications, we could worry about significance. It is clear that Cramer's rule cannot be used for the solution of linear systems of high degree.

1.3 Gaussian Elimination

Gaussian elimination is one of the most commonly used methods for solving linear equations. The procedure may easily be described. We eliminate the last variable between the first and the second equations, between the second and the third and so on. We then have a linear system in one fewer variables. We repeat the elimination procedure until we have one equation in one unknown. The process may now be reversed to determine the other components.

A simple count shows that we have $O(n^3)$ multiplications, a quite feasible number. Thus, there is no difficulty as far as time is concerned.

There are, however, some worrisome features. The elimination procedure involves subtraction and division both of which can lead to loss of significance. Consequently, we are faced with two questions: in what order do we write the equations, and in what order do we eliminate the variables.

Since the time required is so small, it is best to try various orders, since no theory exists at present.

In general, if we have to solve linear systems it is a good idea to use various procedures and compare the results. This is particularly the case if the solution of the linear system occurs in a large calculation.

There is no uniform way of solving linear systems. It is well known, although this has never been established vigorously, that given any procedure there is a system for which the procedure will not give accurate results. The methods we employ depend upon the origin of the linear system and the computational facility we have available.

In general, it is best to avoid linear systems if possible. The theories of dynamic programming and invariant imbedding often allow us to do this.

1.4 On the Evaluation of Determinants

The determinant is a very powerful tool in analysis. In graph theory it is a valuable counting device which can be used to determine the number of graphs of different types. However, the numerical evaluation provides a great deal of difficulty because of the number of operations involved, as mentioned above.

Here we shall present a method which involves of the order of n^5 operations for a determinant of order n. If the associated matrix is positive definite, only n^4 operations are required.

Consider the equation

(1.4.1) $\quad Ax = b.$

Here A is assumed positive definite, and b is a vector with first component one and all the other components zero.

Let us designate the principle minor of A by B. If we solve using Cramer's rule, we have the solution $|B|/|A|$. Here the bars as usual denote the determinant.

Let us now repeat the process with B. If A is nonsingular, B is nonsingular too. We know that the characteristic roots of B separate those of A.

We continue in this fashion. If we now multiply the solutions for the first components, we obtain the desired determinant.

The solution of the linear equation is obtained by the Gaussian method of elimination, or some other method for solving linear systems. This requires of the order of n^3 operations. Hence, the total procedure requires of the order of n^4 operations.

If A is symmetric and nonsingular, we can use A^2 to get a positive definite matrix.

If A is a general matrix, we can use $A'A$ and the well-known relation

(1.4.2) $\quad |A'A| = |A|^2 \quad .$

We see then that the above procedure yields the square of the determinant. If we know the sign of the determinant, we can use these results. If we do not know the sign we have to follow the procedure below.

Let A be a nonsingular matrix. We can then proceed as before. However, the difficulty is that the first component of x may be zero. It is easily seen that all the components of x cannot be zero. Hence, let us take the largest component.

Another way of seeing this is that all the cofactors of the

elements of the first row cannot be zero. If they were zero, the original matrix would be singular.

Thus, we use n^4 operations and the whole procedure involves n^5 operations.

We see then that this is a feasible procedure even for n of the order of magnitude of 100.

1.5 A Basic Formula

Consider the linear system

$$Ax = b. \tag{1.5.1}$$

The solution of Cramer, so useful theoretically, is not useful computationally, as mentioned above. Many methods have been devised for the computational solution. Indeed, this is a good problem because there is no final solution. The technique used depends strongly upon the form of the matrix. Actually, to ensure accuracy, several techniques should be used.

In this chapter, we are interested in selective computation. Often, it is not necessary to determine all the components of x. Let us assume that we want to determine only the first component.

We begin with the formula

$$(x, A^{-1} x) = \max_{y} (2(x,y) - (y,Ay)). \tag{1.5.2}$$

This result may be readily established. Using this result, we can easily obtain various inequalities for the inverse.

Let us call A^{-1}, B. We want to determine an element in B without determining the others. By choosing x suitably, we can easily determine the diagonal elements b_{ii}. Once we have determined these, by a suitable choice of x, we can determine b_{ij}. For a particular i and j, it is necessary to determine b_{ii} and b_{jj}.

To determine the first component of x, we need only the elements b_{1i}.

1.6 Maximization

In this section, we make some remarks about finding the required maximum. To begin with, since the function is concave, we can apply various powerful search methods.

Let us discuss here the "hill-climbing" method. This method is particularly simple in this case because we have a quadratic form. For example, we may fix $y_2,\ldots,y_n$, and then maximize over y_1. This maximization is readily performed explicitly. Having determined the maximum value in this way, we can then pass a plane through a point and continue. Proceeding in this fashion, it is obvious geometrically that we get closer and closer to the maximum. It is now a question of what accuracy is desired. At some point, we can stop this procedure and use a straightforward search procedure.

In many cases, the maximization may be carried out using dynamic programming.

1.7 General Case

We can make the matrix symmetric by using a well-known device

$$A'Ax = A'b. \tag{1.7.1}$$

Here, a prime denotes the transpose of A.

There are many more things to be said.

In the first place, we want to know whether the procedure is done for a single matrix or for many matrices. Secondly, we want to know whether this is carried out for a single matrix or a family of related matrices.

1.8 Green's Function

Let us point out that the basic formula holds for more general operators. With a suitable interpretation of the inner product, we obtain a formula for the Green's function.

The most important case is that where A is a second-order linear differential operator. Consider

$$u'' - q(t)u = f. \tag{1.8.1}$$

We interpret the inner product as follows,

$$(x,y) = \int_0^1 xy\,dt. \tag{1.8.2}$$

Here x and y are one-dimensional functions. If they are vectors, we use the inner product.

A simple integration by parts yields the familiar functional

$$J(u) = \int_0^1 (u'^2 + q(t)u^2 + 2fu)\,dt. \tag{1.8.3}$$

In the multi-dimensional case, the use of Green's Theorem replaces integration by parts.

We can now use methods similar to those above to determine the value of the Green's function at a point.

1.9 Imbedding

Consider the formal identity

$$\operatorname*{Max}_x \left[(x,Ax) - 2(x,y)\right] = (y,A^{-1}y), \tag{1.9.1}$$

where x is an element of a space S, and A is a symmetric operator defined over S with the property that (x,Ax) is negative definite. Since the Euler equation associated with this variational problem is $Ax = y$, we see that the element x furnishing the maximum is given in terms of the inverse operator. For the case of ordinary or partial differential operators, these operators are expressible in terms of Green's functions.

On the other hand, several classes of variational problems of this type can be treated by means of the functional equation technique of dynamic programming. Combining the two approaches, we can derive a number of properties of Green's functions.

Since this approach yields results in a quite straightforward fashion, it is of some interest to see whether or not they can be extended to cover cases in which A is not necessarily negative definite, in which A may be symmetric and complex, and in which it need not be symmetric.

Here we shall show how analytic continuation and a min-max variation can be used to treat the case of a complex symmetric operator, with particular reference to second-order linear differential operators, and how analytic continuation and an imbedding technique can be used to handle nonsymmetric operators.

Let $A + iB$ be a complex symmetric operator. The equation $(A + iB)(x + iy) = u + iv$ reduces to the two real equations

$$Ax - By = u, \qquad Ay + Bx = v. \tag{1.9.2}$$

If A is not negative definite, these can be considered to be the variational equations arising from the problem of determining the maximum over y of the minimum over x of the functional

$$(y,Ay) - (x,Ax) + 2(x,By) + 2(u,x) - 2(v,y). \tag{1.9.3}$$

If A is not negative definite, we replace A by $A - zI$, where z is a sufficiently large positive scalar, and then employ analytic continuation.

As an illustration of this technique, consider the equation

$$u'' + (g(x) + ih(x))u = p(x) + iq(x) \tag{1.9.4}$$

over $a < x < T$. Setting $u = v + iw$, we obtain the equations

$$v'' + g(x)v - h(x)w = p(x),$$

$$w'' + g(x)w + h(x)v = q(x),$$

which are the variational equations connected with the problem of determining the minimum over v and maximum over w of the functional

$$(1.9.5) \qquad \int_a^T [-v'^2 + w'^2 + g(x)v^2 - 2h(x)vw - g(x)w^2$$

$$- 2p(x)v + 2q(x)w]dx.$$

Since this functional is convex in w' and concave in v', it is easy to show that min max = max min.

If $g(x)$ is not uniformly positive in the interval $[a,T]$, we introduce the function $z + g(x)$, where z is a sufficiently large positive quantity and employ analytic continuation.

In a variety of problems in mathematical physics, complex functions occur when energy dissipation is taken into account.

Let A be a nonsymmetric operator, $A \neq A'$. In order to study A^{-1} and the resolvent operator $(A - \lambda I)^{-1}$, by variational techniques, we consider the problem of maximizing the functional

$$(1.9.6) \qquad (x,Bx) + (y,By) + 2(x,Ay) - 2(u,x) - 2(v,y) \quad ,$$

where B is negative definite operator.

The variational equations are

$$(1.9.7) \qquad Bx + Ay = u, \qquad A'x + By = v.$$

We have thus imbedded the equation $Ay = u$, not necessarily of variational origin, within a family of variational equations.

In order to carry out the analytic continuation, we replace B by zB and study the analytic character of the symmetric matrix operator.

$$(1.9.8) \qquad M(z) = \begin{bmatrix} zB & A \\ A' & zB \end{bmatrix}$$

and its inverse, as functions of z. Eventually we wish to set $z = 0$ so as to obtain

$$(1.9.9) \qquad M(0)^{-1} = \begin{bmatrix} 0 & A^{-1} \\ (A')^{-1} & 0 \end{bmatrix}$$

The analytic details in each case will depend upon the nature of the operators A and B.

Bibliography and Comments

Section 1.5. This is a special case of a formula in quasilinearization. See the book, Bellman, R., *Methods of Nonlinear Analysis*, Vol. II, Academic Press, Inc., New York, 1973.

Section 1.6. See the book, Bellman, R., *Introduction to Matrix Analysis*, McGraw-Hill Book Company, New York, 1970.

Section 1.8. For the use of this formula to establish some results for Green's function, see

i. Bellman, R., "On the Non-Negativity of Green's Functions", *Boll. D'Unione Matematico*, Vol. 12, 1957, pp. 411-413.
ii. Bellman, R., "On Variation-Diminishing Properties of Green's Functions", *Boll. D'Unione Matematico*, Vol. 16, 1961, pp. 164-166.

Section 1.9. We are following the paper, Bellman, R., and Lehman, S., "Functional Equations in the Theory of Dynamic Programming. IX. Variational Analysis, Analytic Continuation, and Imbedding of Operators," *Proc. of the National Academy of Sciences*, Vol. 44, No. 9, September 1958, pp. 905-907.

Chapter II

FINDING THE SQUARE ROOT OF A POSITIVE DEFINITE MATRIX

2.1 Introduction

In this chapter, we study the positive definite square root of a positive definite matrix. We are interested in obtaining $(c, A\,c)$. Here c may only assume a finite number of values.

We shall employ two methods. The first is a formula from control theory. The second is a generalization of a well-known algorithm of Hero in a one-dimensional case.

2.2 Use of the Canonical Representation

As is well known, we have

$$A = T' \begin{bmatrix} \lambda_1 & & & & & 0 \\ & \cdot & & & & \\ & & \cdot & & & \\ & & & \lambda_i & & \\ & & & & \cdot & \\ & & & & & \cdot \\ 0 & & & & & \lambda_n \end{bmatrix} \tag{2.2.1}$$

Here the λ_i are the characteristic roots of A. The matrix T has columns characteristic vectors of A. The matrix T' is the adjoint of T.

Using this representation, we can readily form various functions of A.

A disadvantage of this approach is that it requires that we obtain all the characteristic roots and vectors of A.

2.3 A Formula from Control Theory

Let us use the representation

$$(c, A^{1/2}c) = \lim_{T \to \infty} \int_0^T ((u', u') + (u, Au))dt \quad . \tag{2.3.1}$$

This formula may readily be established using the Euler-Lagrange equation.

From this formula, various inequalities for the square root may be obtained.

2.4 Upper and Lower Bounds

Since we have a representation as a minimum, any trial function yields an upper bound. This is the Rayleigh-Ritz method.

Using the adjoint equation, we obtain lower bounds. The adjoint equation will involve the inverse matrix. The methods of Chapter 1 may be used to calculate this inverse.

2.5 Dynamic Programming

Let us write

$$(c, R(Tc)) = \int_0^T ((u', u') + (u, Au))dt \quad . \tag{2.5.1}$$

Then a simple dynamic programming argument yields the Riccati equation

$$R' = A - R^2, \qquad R(0) = 0 \quad . \tag{2.5.2}$$

We would obtain this equation, if we employed the gradient method. It is interesting, however, to see the interpretation.

2.6 Discussion

With various interpretations of the inner product, we have a method of obtaining the square root of general operators.

The determination of the square root of a positive definite matrix plays an important role in the study of two-point boundary value problems, as we shall discuss subsequently.

2.7 Square Roots

The well-known algorithm of Hero is

$$u_{n+1} = \frac{u_n + a/u_n}{2} . \tag{2.7.1}$$

This is an immediate consequence of the Newton-Raphson formula. Analogous relations hold for the higher roots.

We shall use the matrix version of this

$$X_{n+1} = \frac{X_n + AX_n^{-1}}{2} , \quad X_0 = I , \tag{2.7.2}$$

to find the square root of a positive definite matrix A. Analogous formulas hold for other roots, but we are principally interested in the square root because it occurs in variational problems.

2.8 Commutivity

The crucial observation is that all the members of the sequence commute with A. This may easily be proved inductively.

From this, as is well-known, we can reduce all the elements of the sequence to diagonal form with the same transformation. The convergence of the algorithm is thus reduced to the one-dimensional case. As we know, we have quadratic convergence.

2.9 Bounds

We shall use the order relation of positive definite matrices.

The upper bounds are monotone decreasing. If we use the control theory version of the variational problem a single dynamic programming argument yields the differential equation

$$R' = A - R^2 , \qquad R(0) = 0 \quad . \tag{2.9.1}$$

This furnishes a lower bound for each t.

2.10 General Matrices

The same argument can be used for any class of matrices which can be reduced to diagonal form. However, we are not interested in general matrices here since they do not occur in variational problems.

Bibliography and Comments

Section 2.2. See Bellman, R., Introduction to Matrix Analysis, McGraw-Hill Book Company, New York, 1970.

Chapter III

LARGEST AND SMALLEST CHARACTERISTIC ROOTS

3.1 Introduction

In this chapter, we shall investigate the largest characteristic root of a matrix or an integral equation and the smallest characteristic root of the Sturm-Liouville equation.

First, we obtain bounds for the largest characteristic root of a symmetric matrix. Then, using essentially the same method, we obtain bounds for the smallest characteristic root of a Sturm-Liouville equation. Then, we use dynamic programming to obtain an equation for this quantity using the Rayleigh quotient. Next, we turn to integral equations. Here, we shall use a result of Jentzch for nonnegative kernels. Following this, we show that the Green's functions are nonnegative.

In many investigations, it is not necessary to find all the characteristic roots of a matrix. All that is required is the largest one. This is the problem we shall consider here.

First, we shall consider symmetric matrices. In Sec. 3.2, we consider positive definite matrices. In Sec. 3.3, we show how to get better bounds with only a very slight increase in work. In Sec. 3.4, we consider the accuracy of the estimates. In Sec. 3.5, we consider the problem of finding the associated characteristic vector. In Sec. 3.6, we consider the Rayleigh quotient.

In Sec. 3.7, we discuss the case where some of the characteristic roots may be negative.

In Sec. 3.8, we show how these results may be applied to Sturm-Liouville problems. In Sec. 3.9, we give an application to integral equations.

Finally, we discuss the case of a general matrix.

3.2 Positive Definite Matrices

Let A be a positive definite matrix. It follows then that all the characteristic roots are real and positive. Furthermore, we know that the characteristic roots of the powers of A are the corresponding powers of the characteristic roots of A. We also know that the trace, the sum of the elements along the main diagonal, is the sum of the characteristic roots.

Consider the sequence $A, A^2, \ldots, A^n, \ldots$. Let us define

$$u_n = tr(A^n). \tag{3.2.1}$$

Then we have

$$\frac{u_{n+1}}{u_n} \leq \lambda_1 \leq u_n^{1/n} . \tag{3.2.2}$$

Here λ_1 is the largest characteristic value.

A simple application of the Cauchy-Schwarz inequality shows that the lower bound is monotone increasing. A simple result from the theory of inequalities shows that the upper bound is monotone decreasing.

3.3 Better Bounds

Let us now examine the arithmetic. It takes the same labor to multiply two matrices together as to square a matrix. Hence, we consider the sequence $A, A^2, \ldots, A^{2^n} \ldots$. Here, each matrix is the square of the preceding. Let us now define the sequence

$$(3.3.1) \qquad v_n = tr(A^{2^n}) \quad .$$

Then, we have the inequalities

$$(3.3.2) \qquad \left(\frac{v_{n+1}}{v_n}\right)^{1/2^n} \leq \lambda_1 \leq v_n^{1/2^n}$$

As before, the upper bound is monotone decreasing. A simple application of Holder's inequality shows that the lower bound is monotone increasing.

We see that we have to take 2^nth roots. This can be done in several ways. We can use logarithms. Or, we can take repeated square roots. Here, we can use the simple recoverance relation of Hero and obtain arbitrary accuracy. Even if we have a computer with a limited number of significant figures, a simple use of algebra can overcome this.

3.4 Accuracy of Estimates

Let us now discuss the accuracy of the estimates.

Let λ_1 denote the largest characteristic value, and λ_2 denote the next largest. The accuracy of the bounds depends upon the ratio λ_1 to λ_2. We shall obtain estimates for this ratio by obtaining estimates for $\lambda_1\lambda_2$.

Consider the elementary sum $\sum_{i \neq j} \lambda_i\lambda_j$. We can obtain an expression for this directly from the matrix. However, it is easier to proceed as follows. We have the elementary identity

$$(3.4.1) \qquad 2 \sum_{i \neq j} \lambda_i\lambda_j = (\textstyle\sum_i \lambda_i)^2 - \sum_i \lambda_i{}^2 \quad .$$

This is equal to

(3.4.2) $\quad (\operatorname{tr} A)^2 - \operatorname{tr}(A^2).$

We can now proceed as above and obtain similar bounds.

3.5 Associated Characteristic Vector

It is sometimes desirable to obtain the associated characteristic vector. We know the components of this vector can be taken to be the co-factors of the elements of the first row. The procedure for evaluating determinants given in Chapter 1, can be useful.

3.6 The Rayleigh Quotient

We know that we have

$$(3.6.1) \qquad \lambda_1 = \max_x \frac{(x, Ax)}{(x,x)} .$$

Any choice of x yields a lower bound. This procedure, however, has several drawbacks. In the first place, we don't know how accurate this lower bound is. In the second place, we possess no systematic way of improving the bound.

3.7 Symmetric Matrices

Let us still consider symmetric matrices, which are no longer necessarily definite. We can overcome this difficulty by using the matrix e^A. That method, given in Chapter 4, can be used to calculate the exponential matrix.

3.8 Application to Sturm-Liouville Equations

The method given above can be applied to the Sturm-Liouville equation,

$$(3.8.1) \qquad u'' - \lambda k(x)u = 0, \qquad u(0) = u(\pi) = 0 \quad ,$$

if we use a finite difference approximation. The problem of

determining the smallest characteristic root of the Sturm-Liouville equation has now been approximated by the problem of obtaining the largest characteristic root of a positive definite matrix.

In a later section we will apply dynamic programming to a treatment of the Sturm-Liouville equation using the continuous analog of the Rayleigh quotient.

3.9 Application to Integral Equations

Consider the integral equation

$$\lambda f(x) = \int_0^1 k(x,y)f(y)dy \quad . \tag{3.9.1}$$

We now can proceed in two ways. The formulas given above have analogs for integral equation. Alternatively, we can use a quadrature method and reduce the solution of (3.9.1) to a matrix equation.

3.10 General Matrices

Let us now consider the problem of finding the root of greatest amplitude of a general matrix. This root in general is complex. Thus, for large n we have

$$tr(A^n) \cong (\lambda + iv)^n + (\lambda - iv)^n \quad . \tag{3.10.1}$$

If we combine this relation with the relation for $2n$, we can easily determine λ and v.

We no longer have upper and lower bounds. There is no question that the determination of the characteristic root of largest amplitude for a general matrix is quite difficult.

3.11 The Smallest Characteristic Root of a Sturm-Liouville Equation

In this section, we are interested in the problem of determining the characteristic values of the Sturm-Liouville equation

3.11.1 $u'' + \lambda a(t)u = 0$, $u(0) = u(1) = 0$.

It will be clear from what follows that the methods we discuss can be applied to questions of this type involving quite general boundary conditions, as long as the interval is finite.

There are, at present, a number of powerful techniques available for treating problems of this genre, based upon variational techniques, and upon matrix techniques applied to finite difference version of the foregoing differential equation.

The variational approach depends upon the fact that if $a(t)$ satisfies a reasonable condition such as

$$(3.11.2) \qquad 0 < a^2 \leq a(t) \leq b^2 < \infty , \qquad 0 \leq t \leq 1 ,$$

then the characteristic values, $\lambda_1 < \lambda_2 < \ldots$, are the respective relative minima of the functional

$$(3.11.3) \qquad J(u) = \int_0^1 u'^2 dt / \int_0^1 a(t)u^2 dt$$

as u ranges over the space of functions for which the integrals exist and for which $u(0) = u(1) = 0$.

In particular,

$$(3.11.4) \qquad \lambda_1 \leqq \int_0^1 u'^2 dt / \int_0^1 a(t)u^2 dt$$

for all functions $u(t)$ satisfying the prescribed boundary conditions. We thus have a means of obtaining upper bounds for λ_1 which turn out to be remarkably accurate even for simple choices of trial functions $u(t)$.

Another method is based upon using equations of the form

$$(3.11.5) \qquad u_{n+2} - 2u_{n+1} + u_n + \lambda\Delta^2 a_n u_n = 0 ,$$

$u(0) = u(N) = 0$, and applying any of a number of methods used to derive the characteristic roots and vectors of a symmetric matrix.

There is, however, a significant difference between a problem of this type, and the Sturm-Liouville problem described above. This is due to the fact that it is quite easy to find asymptotic solutions to (3.11.1) for large λ, and thus, approximate expressions for the higher characteristic values.

Let, for simplicity of notation, $a(t) = q^2(t)$; then the Liouville transformation $s = \int_0^t q(t_1)dt_1$, converts

$$u'' + \lambda q^2(t)u = 0 \tag{3.11.6}$$

into

$$\frac{d^2u}{ds^2} + \frac{q'(t)}{q^2(t)}\frac{du}{ds} + \lambda u = 0 \quad . \tag{3.11.7}$$

The further transformation

$$v = u\sqrt{q(t)} = u \exp\left\{\frac{1}{2}\int \frac{q'(t)}{q^2(t)} ds\right\} \tag{3.11.8}$$

converts (3.11.7) into

$$\frac{d^2v}{ds^2} + \left[\lambda - \frac{1}{2}\frac{d}{ds}\left(\frac{a'(t)}{a^2(t)}\right) - \frac{1}{4}\left(\frac{a'(t)}{a^2(t)}\right)^2\right]v = 0 \quad . \tag{3.11.9}$$

The new boundary conditions are

$$v(0) = 0 \ , \qquad v\left(\int_0^1 q(t)dt\right) = 0 \quad . \tag{3.11.10}$$

Writing (3.11.9) in the form

$$v''(s) + (\lambda + b(s))v(s) = 0 \quad , \tag{3.11.11}$$

we know that we can find asymptotic developments for $v(s)$ starting from the integral equation

$$(3.11.12) \qquad v(s) = c_1 \cos \lambda^{1/2}s + c_2 \sin \lambda^{1/2}s - \int_0^s \frac{[\sin \lambda^{1/2}(s-r)]}{\lambda^{1/2}} b(r)v(r)dr$$

and iterating. Approximate values of λ are now determined by means of the constraint $v\left(\int_0^1 q(t)dt\right) = 0$. Thus, the higher characteristic values have the principle term

$$(3.11.13) \qquad \lambda_n \cong n^2\pi^2 \Big/ \left(\int_0^1 q(t)dt\right)^2 \quad .$$

To obtain more precise results, we can find further terms of the asymptotic series derived from (3.11.12), and we can combine this with numerical integration of (3.11.1).

It follows from these considerations that the greatest difficulty is experienced in obtaining accurate estimations of the first characteristic value. In many investigations this is all that is desired.

We wish to present a new method, suitable for hand digital computer calculation, which furnishes monotone convergence, through sequences of upper and lower bounds, to the smallest characteristic value. Similar sequences can be used to obtain monotone convergence to products of the form $\prod_{i}^{k} = \lambda_{i} = 1$. The method has the advantage of permitting λ_1 to be determined to a high degree of accuracy.

To illustrate these techniques, we consider the equation

$$(3.11.14) \qquad u'' + \lambda(1 + t)u = 0 \ , \qquad u(0) = u(1) = 0 \ ,$$

which is connected with Airy's function, or Bessel functions of order $\frac{1}{3}$.

3.12 The Equation Determining the Characteristic Values

Let us note in passing that the method we use is an application of an approach we have used, in various lecture courses on differential equations, to derive the fundamental results of Sturm-Liouville theory.

Consider the linear differential equation

$$u'' + \lambda a(t)u = 0 , \quad u(0) = 0 , \quad u'(0) = 1 . \tag{3.12.1}$$

The solution of this initial value problem may be obtained over $0 \le t \le 1$ as a power series in λ in the form

$$u = t + \sum_{n=1}^{\infty} u_n(t) \lambda^n , \tag{3.12.2}$$

where the sequence of coefficient functions $\{u_n(t)\}$, $n = 1,2,...$ may be determined by recurrence relations

$$u_0(t) = t, \tag{3.12.3}$$

$$u_n(t) = - \int_0^t (t - s)u_{n-1}(s)a(s)ds , \quad n = 1,2,...$$

It is easy to see that u, as defined by (3.12.2), is an analytic function of λ for all finite λ for $0 \le t \le 1$, The roots of the equation

$$f(\lambda) = u(1) = 1 + \sum_{n=1}^{\infty} u_n(1)\lambda^n = 0 \tag{3.12.4}$$

are the desired characteristic values.

3.13 Discussion

If we assume that the sequence of coefficients is determined by means of either a hand or machine computation, a matter we will discuss again below, there is a problem of determining the first few roots of the equation in (3.12.4).

This is a problem which can be treated in several ways. It would seem that an efficient procedure would be to use the sequences we shall describe presently to obtain reasonably accurate estimates for the characteristic values, and then use a Newton-Raphson method, or a modification, to obtain very accurate values.

3.14 Analytic Preliminaries

Referring to the equation in (3.12.2), we easily see that

(3.14.1) $$|u(t)| \le c^{k|\lambda|^{1/2}} ,$$

for $0 \le t \le 1$, where k is a constant. Consequently, the Weierstrass factorization of $f(\lambda)$ takes the form

(3.14.2) $$f(\lambda) = \prod_{i=1}^{\infty} (1 - \lambda/\lambda_i) .$$

As we know, $\lambda_n = O(n^2)$ as $n \to \infty$, in view of the assumptions we have made concerning $a(t)$ in (3.11.2).

Our aim is now, by following the technique used by Newton to relate the sums of the powers of the roots and the elementary symmetric functions, which are the coefficients, to obtain relations for the sums

(3.14.3) $$b_r = \sum_{i=1}^{\infty} 1/\lambda_i^r , \qquad r = 1,2,\ldots$$

in terms of the coefficents $u_n(1)$.

It is clear that

$$(3.14.4) \qquad \log f(\lambda) = \sum_{i=1}^{\infty} \log (1 - \lambda/\lambda_i) = - \sum_{r=1}^{\infty} (\lambda^r/r) \cdot \left\{ \sum_{i=1}^{\infty} 1/\lambda_i \right\} = - \sum_{r=1}^{\infty} \lambda^r b_r/r \quad ,$$

for $|\lambda| < \lambda_1$.

It is important then to obtain the coefficients of the expression of $\log f(\lambda)$. Although this can be done directly, it is easier to proceed as follows. Write

$$(3.14.5) \qquad \log f(\lambda) = \sum_{k=1}^{\infty} c_k \lambda^k \quad .$$

Then

$$(3.14.6) \qquad f'(\lambda)/f(\lambda) = \sum_{k=1}^{\infty} kc_k \lambda^{k-1} \quad ,$$

whence

$$(3.14.7) \qquad \sum_{n=1}^{\infty} nu_n(1)\lambda^{n-1} = \left(\sum_{k=1}^{\infty} kc_k \lambda^{k-1} \right) \left(1 + \sum_{n=1}^{\infty} u_n(1)\lambda^n \right) \quad ,$$

whence we obtain the well-known recurrence relations

$$(3.14.8) \qquad nu_n = nc_n + \sum_{k=1}^{n-1} kc_k u_{n-k} \quad .$$

These permit us to calculate the c_n in a very simple fashion once the sequence $\{u_r(1)\}$ has been determined, and thence the b_n.

3.15 Inequalities

Let us now show that the sequence $\{b_k\}$ can be used to obtain sequences which converge monotonically from above and below to the first characteristic value λ_i.

THEOREM 1. We have the inequalities

$$(3.15.1) \qquad b_k/b_{k+1} > \lambda_1 > 1/b_k^{1/k} , \qquad k = 1,2,\ldots .$$

The sequence $\{b_k/b_{k+1}\}$ is monotone decreasing; the sequence $\{1/b_k^{1/k}\}$ is monotone increasing, and

$$(3.15.2) \qquad \lambda_1 = \lim_{k\to\infty} b_k/b_{k+1} = \lim_{k\to\infty} 1/b_k^{1/k} .$$

PROOF. The monotone character of the ratio b_k/b_{k+1} follows directly from Schwarz's inequality, since

$$(3.15.3) \qquad b_k^2 = \left(\sum_{i=1}^{\infty} 1/\lambda_i^k\right)^2 = \left(\sum_{i=1}^{\infty} 1/\lambda_i^{(k+1)/2}\,\lambda_i^{(k-1)/2}\right)^2$$

$$\leq \left(\sum_{i=1}^{\infty} 1/\lambda_i^{k+1}\right)\left(\sum_{i=1}^{\infty} 1/\lambda_i^{k-1}\right) = b_{k+1}\,b_{k-1} .$$

The monotone behavior of $b_k^{1/k}$ is a consequence of the well-known inequality

$$(3.15.4) \qquad \left(\sum_{i=1}^{\infty} x_i\right) \geq \left(\sum_{i=1}^{\infty} x_i^2\right)^{1/2} \geq \left(\sum_{i=1}^{\infty} x_i^3\right)^{1/3} \geq \cdots ,$$

for any set of non-negative x_i.

The proof of the limiting relation is clear.

3.16 Rate of Convergence

Since

$$(3.16.1) \qquad \frac{b_k}{b_{k+1}} = \frac{(1/\lambda_1^k)\,[1 + (\lambda_1/\lambda_2)^k + \ldots]}{(1/\lambda_1^{k+1})\,[1 + (\lambda_1/\lambda_2)^{k+1} + \ldots]}$$

$$= \lambda_1 \left\{1 + (\lambda_1/\lambda_2)^k - (\lambda_1/\lambda_2)^{k+1} + \ldots\right\} ,$$

we see that

$$(3.16.2) \qquad b_k/b_{k+1} - \lambda_1 \cong \lambda_1(\lambda_1/\lambda_2)^k$$

for large k.

Similarly,

$$(3.16.3) \qquad b_k^{1/k} = (1/\lambda_1)\,(1 + (\lambda_1/\lambda_2)^k + \ldots)^{1/k}$$

$$\cong (1/\lambda_1)\,(1 + k(\lambda_1/\lambda_2)^k)$$

for large k.

It is to be expected that b_k/b_{k+1} will furnish a better approximation to λ_1 for large k.

3.17 Example

Let us consider the case where $a(t) \equiv 1$, $\lambda_1/\lambda_2 = \frac{1}{4}$. Consequently, in general, the rate of convergence of these sequences will not be too rapid. There are two things we can do to obtain more accurate estimations of λ_1. In the first place, we can use the root-squaring technique. Since

$$(3.17.1) \qquad f(\lambda) = \prod_{i=1}^{\infty} (1 - \lambda/\lambda_i) \quad .$$

we see that

$$(3.17.2) \qquad f_1(\lambda) = f(\lambda^{1/2})\, f(-\lambda^{1/2}) = \prod_{i=1}^{\infty} (1 - \lambda/\lambda_i^2) \quad .$$

Using the power series development for $f_1(\lambda)$ we obtain a sequence $\{b_k'\}$ with

$$(3.17.3) \qquad \lim_{k\to\infty} b_k'/b_{k+1}' = \lambda_1^2 \quad ,$$

and a rate of convergence depending upon $(\lambda_1/\lambda_2)^2$.

Alternatively, once we have an estimate for λ_1 with an accuracy of 1 in 10^{-s}, we can then turn to the power series for $f(\lambda)$ and use the Newton approximation technique,

$$\lambda_1^{(n+1)} = \lambda_1^{(n)} - f(\lambda_1^{(n)})/f'(\lambda_1^{(n)}) \quad . \tag{3.17.4}$$

This will yield a further approximation with accuracy of essentially 1 in 10^{-2s}. Continued use of this technique is limited only by the number of $u_n(1)$ which are computed, and the accuracy of this computation. There is no difficulty involved in using this technique here, since we know from theoretical considerations that the roots of $f(\lambda)$ are simple.

3.18 Inequalities for $\prod_{i=1}^{R+1} \lambda_i$

Similar upper bounds can be obtained for the products $\prod_{i=1}^{R+1} \lambda_i$, $R = 1,2,\ldots$.

Consider the determinant

$$b_k^{(R)} = \begin{vmatrix} b_k & b_{k+1} & \cdots & b_{k+R} \\ \cdot & \cdot & \cdots & \cdot \\ \cdot & \cdot & \cdots & \cdot \\ \cdot & \cdot & \cdots & \cdot \\ b_{k+R} & b_{k+R+1} & \cdots & b_{k+2R} \end{vmatrix} \quad , \quad R = 1,2,\ldots \tag{3.18.1}$$

It is not difficult to show that

$$\lim_{k\to\infty} b_k^{(R)}/b_{k+1}^{(R)} = \lim_{k\to\infty} (b_k^{(R)})^{-1/k} = \lambda_1\lambda_2\cdots\lambda_{R+1} \quad . \tag{3.18.2}$$

To show that

$$b_k^{(R)}/b_{k+1}^{(R)} > b_{k+1}^{(R)}/b_{k+2}^{(R)} \quad , \quad k = 1,2,\ldots \tag{3.18.3}$$

for $R = 1,2,\ldots$ we use the well-known fact that the matrix

$$B_k^{(R)} = \begin{pmatrix} b_k & b_{k+1} & \cdots & b_{k+R} \\ \cdot & & \cdots & \cdot \\ \cdot & & \cdots & \cdot \\ \cdot & & \cdots & \cdot \\ b_{k+R} & b_{k+R+1} & \cdots & b_{k+2R} \end{pmatrix} \tag{3.18.4}$$

is positive definite for all k and R, and hence that $(B_k^{(R)})^{-1}$ is positive definite.

The sequence $(b_k^{(R)})^{-1/k}$ does not seem to have any simple monotonicity properties.

3.19 The Equation $u'' + \lambda(1 + t)u = 0$

Let us now illustrate some of the ideas discussed above by means of the equation

$$u'' + \lambda(1 + t)u = 0 , \qquad u(0) = u(1) = 0 \quad . \tag{3.19.1}$$

The first problem we face is that of computing the sequence $\{u_n(t)\}$ by means of the recurrence relations of (3.12.3). Since $u(t)$ is an entire function of λ for $0 \leq t \leq 1$, the coefficients, $u_n(t)$, become quite small as n increases. If $a(t) \equiv 1$, the coefficient of λ^n is $(-1)^n/(2n + 1)!$ Hence, if we are using a digital computer, even one with floating point arithmetic, it is necessary to renormalize. A very simple renormalization is one which sets

$$v_n(t) = (-1)^n \, (2n + 1)! \, u_n^{(t)} \quad . \tag{3.19.2}$$

Then

$$v_n(t) = \frac{1}{(2n + 1)\, 2n} \int_0^t (t - s)\, v_{n-1}(s) a(s) ds \, , \qquad n = 1,2,\ldots \tag{3.19.3}$$

$$v_0(t) = 1 \quad .$$

Since (3.19.3) is equivalent to the differential recurrence relation

(3.19.4) $\quad v_n''(t) = a(t)\, v_{n-1}(t)/2n\,(2n+1) \qquad v_n(0) = v_n'(0) = 0$,

we can use a Runge-Kutta integration procedure to obtain fairly accurate values of $v_n(1)$ (See Table 3.1).

Table 3.1

n	$v_n(1) = (-1)^n\,(2n+1)!\,u_n(1)$
0	1. 000 000 000
1	1. 499 999 92
2	2. 238 094 66
3	3. 333 330 15
4	4. 960 358 93
5	7. 378 146 87
6	10. 971 261 4
7	16. 310 824 0
8	24. 244 529 3
9	36. 028 967 6
10	53. 522 379 4

The decision as to how many elements of the sequence $\{u_n(1)\}$ to compute depends upon an *a priori* estimate of the magnitude of λ_1, the time involved in the computation, the accuracy of the computation, and the accuracy with which it is desired.

Since $1 + t \geq 1$, we see that $\lambda_1 < \pi^2 < 10$. Hence the order of magnitude of the last term computed in the power series would be

$$(3.19.5) \qquad u_n(1)\,\lambda_1^n < \frac{53.5}{(21)!}\,10^{10} < \frac{10^2 \cdot 10^{10}}{(20)!} < \frac{10^2 \cdot 10^{10}}{2^{20} \cdot 10^{20}} = \frac{10^{-8}}{2^{20}}$$

(using Sterling's approximation). This is more than sufficient, considering the inaccuracy involved in numerical integration, for the determination of λ_1, and is sufficient for the determination $\lambda_2 \leq 4\pi^2$.

The next step is to compute the sequence of coefficients in $\log f(\lambda)$, namely $\{b_k\}$, using (3.14.8). The results are given in Table 3.2, together with the ratios b_k/b_{k+1} and the roots $b_k^{-1/k}$.

Table 3.2

k	b_k	b_k/b_{k+1}	$b_k^{-1/k}$ (slide rule evaluation)
1	25.0000	9.921 26 x 10^{-2}	4.00 x 10^{-2}
2	251.984	6.958 90 x 10^{-2}	6.30 x 10^{-2}
3	3621.03	6.632 47 x 10^{-2}	6.51 x 10^{-2}
4	54595.5	6.567 79 x 10^{-2}	6.54 x 10^{-2}
5	831261.0	6.553 06 x 10^{-2}	6.55 x 10^{-2}
6	12685100.0	6.549 54 x 10^{-2}	-
7	193679 x 10^3	6.548 66 x 10^{-2}	-
8	29575 x 10^5	- -	-

For the purpose of using the Newton-Raphson scheme mentioned above, (3.17.4), we see that b_4/b_5 and $b_5^{-1/4}$ yield sufficiently good initial approximations with an error of about 1 in 600. One or two applications of (3.17.4) would yield λ_1 to an accuracy sufficient for most purposes.

The convergence of the sequence for $\lambda_1 \lambda_2$ is much less rapid, as is to be expected. The results are shown in Table 3.3.

Table 3.3

k	$b_k^{(1)} = b_k b_{k+2} - b_{k+1}^2$	$b_k^{(1)}/b_{k+1}^{(1)}$	$(b_k^{(1)})^{-1/k}$
1	27030.0	418.85×10^{-4}	$.37 \times 10^{-4}$
2	645330.0	219.85×10^{-4}	12.45×10^{-4}
3	29353×10^3	188.93×10^{-4}	32.45×10^{-4}
4	15537×10^5	179.34×10^{-4}	50.37×10^{-4}
5	86634×10^6	-	-

Using the value of λ_1 obtained above, we obtain a first approximation of $\lambda_2 \cong 27$. From the monotonicity of the ratios, we know that λ_2 is actually less than this. An application of Newton's approximation will yield a greatly improved result.

Note that λ_2 is sufficiently large so that the asymptotic techniques discussed in Sec. 3.11 can be used to provide an independent check of the accuracy of the first approximation to λ_2.

3.20 Alternate Computational Scheme for Polynomial Coefficients

In what has preceded, we have spoken in terms of numerical evaluation of the sequence $\{u_n(t)\}$. Although this procedure has the great advantage of straightforwardness and simplicity, via hand computation or digital computation, it suffers from the fact that errors of integration arise, and grow with each new member of the sequence.

Consequently, it is worth noting a special, but important, case in which we can avoid mechanical quadrature and carry out the entire operation by hand.

Suppose that $a(t)$ is a polynomial of the form

(3.20.1) $$a(t) = a_0 + a_1 t + \dots + a_k t^k \quad .$$

It will be clear then that the elements of the sequence $\{u_n(t)\}$ will also be polynomials. Furthermore, it is clear that $u_n(t)$ will have the form

(3.20.2) $$u_n(t) = a_0 t^{2n+1}/(2n+1)! + \dots + a_{kn} t^{2n+k} + \dots \quad .$$

Using the recurrence relation of (3.12.3), we can then obtain linear recurrence relations for the sequence $\{a_{kn}\}$, $k = 1,2,\dots; \ n = 1,2,\dots$.

There are a number of renormalization questions concerned with the effective calculation of the sequence, and asymptotic relations which can be used to speed the computation. A discussion of these would take us too far afield.

3.21 Extension to Higher Order Equations

Let us now consider the equation

(3.21.1) $$u^{(4)} + \lambda a(t) u = 0$$

with the boundary conditions

(3.21.2) $$u(0) = u'(0) = 0 . \qquad u(1) = u'(1) = 0 \quad .$$

Proceeding as above, we consider the solution, $u(t,\lambda)$, of the initial value problem

(3.21.3) $$u(0) = 0 , \qquad u'(0) = 0 , \qquad u''(0) = c_1 ,$$
$$u'''(0) = c_2 ,$$

which we can write in the form

$$u = c_1\, u_1(t,\lambda) + c_2\, u_2(t,\lambda) \quad , \tag{3.21.4}$$

where u_1 and u_2 are determined by the initial conditions

$$\begin{aligned} &u_1(0) = 0\ , \quad u_2(0) = 0\ , \quad u_1'(0) = 0\ , \quad u_2'(0) = 0\ , \\ &u_1''(0) = 1\ , \quad u_2''(0) = 0\ , \quad u_1'''(0) = 0, \quad u_2'''(0) = 1\ . \end{aligned} \tag{3.21.5}$$

As before, there is no difficulty in obtaining the power series developments in terms of λ for the functions u_1 and u_2.

Applying the boundary conditions in (3.21.2), we obtain the simultaneous equations

$$c_1\, u_1(1,\lambda) + c_2\, u_2(1,\lambda) = 0, \quad c_1\, u_1'(1,\lambda) + c_2\, u_2'(1,\lambda) = 0 \tag{3.21.6}$$

whence the determining equation for λ is

$$f(\lambda) = \begin{vmatrix} u_1(1,\lambda) & u_2(1,\lambda) \\ u_1'(1,\lambda) & u_2'(1,\lambda) \end{vmatrix} = 0 \quad . \tag{3.21.7}$$

From here on, the argumentation is as before.

3.22 The Rayleigh Quotient and Dynamic Programming

Consider the equation

$$u'' + \lambda a(t) u = 0 \tag{3.22.1}$$

where we have the boundary conditions

$$u(0) = u(T) = 0 \quad . \tag{3.22.2}$$

We shall assume that $a(t)$ is positive throughout the interval and possesses a Taylor expansion in the neighborhood of zero. We know that the characteristic values are real, positive, and simple.

The smallest characterisitic value is given by the Rayleigh quotient

$$(3.22.3) \qquad \lambda = \min_u \frac{\int_0^T u'^2 dt}{\int_0^T a(t) u^2 dt} ,$$

where u is subject to the same boundary conditions as above.

The purpose of these sections is to derive a nonlinear partial differential equation for which λ is one value of the solution. In Sec. 3.23, we derive this equation using a straightforward dynamic programming approach. In Sec. 3.24, we discuss some computational aspects of determining the solution of this equation. In Sec. 3.25, we show that the same method may be applied to the nonlinear characteristic value problem. In Sec. 3.26, we discuss how the method may be applied to find the higher characteristic values. In Sec. 3.27, we discuss how the same method may be applied to some matrix problems. Finally, in Sec. 3.28, we discuss selective computation.

3.23 Dynamic Programming Approach

Let us consider the more general problem

$$(3.23.1) \qquad f(c,T) = \min_u \int_0^T u'^2 dt$$

where u is subject to

$$(3.23.2a) \qquad u(T) = c , \qquad u(0) = 0 ,$$

$$(3.23.2b) \qquad \int_0^T a(t) u^2 dt = 1 .$$

We now make the decomposition

$$\int_0^T = \int_0^{T-\Delta} + \int_{T-\Delta}^T \tag{3.23.3}$$

In the second integral we make the renormalization

$$u = (1 - c^2 \Delta/2)v \quad . \tag{3.23.4}$$

We thus obtain, to terms in Δ^2,

$$f(c,T) = v^2 + (1 - c^2)\, f(c + v,\, T - \Delta) \quad . \tag{3.23.5}$$

Passing to the limit we obtain the nonlinear partial differential equation

$$f_T = - cf - f_c^2/4 \quad . \tag{3.23.6}$$

3.24 Computational Aspects

The nonlinear partial differential equation above cannot be solved routinely on a digital computer. In the first place, we see that the solution becomes unbounded as $T \to 0$. In the second place, we see that there is a singularity as both c and $T \to 0$.

To obtain the numerical solution, we can use analysis to derive the solution for small T. This is consistent with the general principle that effective computing needs accommodation of analysis and the arithmetic capabilities of the digital computer. We also use analysis to derive the solution for small c and T. If we keep the constant term in the Taylor expansion of $a(t)$, we can solve the associated differential equation in terms of trigonometric functions; if we keep the first two terms we can solve the equation in terms of Bessel functions of order 1/3. If we keep the first three terms, we can solve the differential equation in terms of cylinder functions. Since we have to use power series expansions at some stage, it is probably best to avoid special functions and carry through the whole calculation using power series.

3.25 Nonlinear Characteristic Value Problems

Let us consider the function of three variables defined by

$$f(c_1, c_2, T) = \min_u \int_0^T u'^2 dt \tag{3.25.1}$$

where u is subject to

$$u(T) = c_1 , \quad u(0) = 0 \tag{3.25.2a}$$

$$\int_0^T g(u)dt = c_2 \quad . \tag{3.25.2b}$$

If g is a power of u, we can use renormalization as above to eliminate the variable c_2. If we have

$$g(u) = u^2 + au^n , \tag{3.25.3}$$

we can use renormalization and take a as a state variable.

The same method may be applied to general isoperimetric problems.

3.26 Higher Characteristic Values

Let us now turn to the determination of higher characteristic values. It will be sufficient to consider the second characteristic value. We know that there is orthogonality. Hence, we add the condition

$$\int_0^T uu_1 dt = c_2 \quad . \tag{3.26.1}$$

Here, u_1 is the first characteristic function. The dynamic programming approach given above determines the function too.

We can now proceed as above to obtain a nonlinear partial differential equation.

3.27 Matrix Theory

If we have a symmetric matrix, we can obtain Rayleigh

quotients for the smallest and largest characteristic values.

If the matrix has a particular structure, we can use a dynamic programming approach.

Matrix problems are obtained if the Ritz-Galerkin method is used.

In many cases, the matrix is non-negative as well as symmetric. We can obtain upper and lower bounds from this fact; see below.

3.28 Selective Computation

In many cases, we only want a few values of the nonlinear partial differential equation. In that case, we can solve the associated variational problem instead.

3.29 On the Integral Equation $\lambda f(x) = \int_0^a K(x-y)\, f(y)dy$

We wish to consider the integral equation

$$\lambda f(x) = \int_0^a K(x-y)\, f(y)dy \;, \qquad a > 0 \;, \tag{3.29.1}$$

which occurs in connection with various problems of probability theory and mathematical physics. Unless $K(x)$ is a function of particularly simple type, such as polynomial or sum of exponentials, the problem of obtaining an exact solution of (3.29.1) appears exceedingly difficult. In the present sections we discuss the behavior of the largest characteristic value, λ_M, as $a \to \infty$, under certain assumptions concerning $K(x)$, and illustrate our results with reference to the integral equation of Kac,

$$\lambda f(x) = \int_0^a e^{-(x-y)^2}\, f(y)dy \;. \tag{3.29.2}$$

The principal result is

THEOREM 1. If

(a) $K(x)$ is non-negative, even, and monotone decreasing for $0 \leq x < \infty$,

(3.29.3) (b) $$c = \int_0^\infty K(x)dx < \infty ,$$

then as $a \to \infty$, $\lambda_M \to 2c$.

More precisely, for all $a \geq 0$,

$$2\int_0^{a/2} K(x)dx \geq \lambda_M \geq 2\int_0^a K(x)dx - \frac{2}{a}\int_0^a xK(x)dx \quad . \tag{3.29.4}$$

Our first method of proof depends upon two tools, the classical Rayleigh-Ritz procedure and a new variational procedure introduced by Bohnenblust. The second method utilizes some known techniques of the theory of integral equations, and exhibits an important property of the characteristic function associated with λ_M.

3.30 First Proof

We shall employ the following two lemmas, the first of which is well-known:

LEMMA 1. If $K(x,y)$ is real, symmetric, and satisfies the condition that

$$\int_0^a\int_0^a K^2(x,y)dxdy < \infty ,$$

then

$$\lambda_M = \operatorname*{Max}_f \frac{\int_0^a\int_0^a K(x,y)f(x)f(y)dxdy}{\int_0^a f^2(x)dx} \quad . \tag{3.30.1}$$

LEMMA 2. If $K(x,y)$ is bounded and non-negative for $0 \leq x, y \leq a$, and λ_M denotes, as above, the largest characteristic value of $K(x,y)$, then

$$\text{(3.30.2)} \quad \begin{aligned} \operatorname*{Sup}_{g \geq 1} \operatorname*{Min}_{x} \frac{\int_0^a K(y,x)g(y)dy}{g(x)} &\leq \lambda_M \\ &\leq \operatorname*{Inf}_{g \geq 1} \operatorname*{Max}_{x} \frac{\int_0^a K(y,x)g(y)dy}{g(x)} . \end{aligned}$$

PROOF OF LEMMA 2. As is known, the characteristic function associated with λ_M may be taken to be positive, by virtue of the non-negativity of $K(x,y)$, taking K to be nontrivial. Let $g(x)$ be a positive function greater than or equal to one. From

$$\text{(3.30.3)} \quad \lambda_M f(x) = \int_0^a K(x,y)f(y)dy ,$$

we obtain

$$\text{(3.30.4)} \quad \begin{aligned} \lambda_M \int_0^a f(x)g(x)dx &= \int_0^a \left(\int_0^a K(x,y)g(x)dx \right) f(y)dy \\ &= \int_0^a \frac{\left(\int_0^a K(x,y)g(x)dx \right)}{g(y)} f(y)g(y)dy \end{aligned}$$

whence (3.30.2) follows immediately. That the two sides of the inequality in (3.30.2) are actually equal and equal to λ_M is a result of Bohnenblust.

Lemma 1 contains the essence of the Rayleigh-Ritz method and furnishes lower bounds for λ_M. Lemma 2, which is also based upon variational principles, furnishes upper and lower bounds. Combining the two, and using the fact that $K(x)$ is even, we obtain

$$\underset{g \geq 1}{\text{Inf}} \ \underset{0 \leq x \leq a}{\text{Max}} \ \frac{\int_0^a K(x - y)g(y)dy}{g(x)} \geq \lambda_M$$

(3.30.5)
$$= \underset{f}{\text{Max}} \ \frac{\int_0^a \int_0^a K(x - y)f(x)f(y)dxdy}{\int_0^a f^2 \ (x)dx}$$

$$\leq \underset{g \geq 1}{\text{Sup}} \ \underset{0 \leq x \leq a}{\text{Min}} \ \frac{\int_0^a K(x - y)g(y)dy}{g(x)} .$$

The simplest possible choices of f and g, viz., $f = g = 1$, yield (3.29.4). It is clear that these results may be further refined by a cleverer choice of f and g. However, the calculations rapidly become complicated.

Setting $f = 1$, we obtain

$$\lambda_M \geq \int_0^a \left[\int_0^a K(x - y)dy \right] dx/a$$

(3.30.6)
$$= \frac{1}{a} \int_0^a \left[\int_0^x K(u)du + \int_0^{a-x} K(u)du \right] dx$$

$$= \frac{2}{a} \int_0^a \left[\int_0^x K(u)du \right] dx \quad .$$

Integration by parts yield

(3.30.7)
$$\lambda_M \geq 2 \int_0^a K(u)du - \frac{2}{a} \int_0^a uK(u)du \quad .$$

Setting $g = 1$, we obtain

$$\text{(3.30.8)} \qquad \max_{0 < x < a} \int_0^a K(x - y)dy \geq \lambda_M \quad .$$

Since K is even and monotone decreasing, it is easily seen that the maximum occurs at $x = \frac{a}{2}$. Thus,

$$\text{(3.30.9)} \qquad \int_0^a K\left(\frac{a}{2} - y\right) dy = 2 \int_0^{a/2} K(y)dy \geq \lambda_M \quad .$$

If $\int^\infty K(x)dx < \infty$, it follows readily that $\int_0^a xK(x)dx = O(a)$ as $a \to \infty$, and thus $\lambda_M \to 2 \int_0^\infty K(u)du$ as $a \to \infty$.

The bounds for λ_M obtained in this way will be narrow only for fairly large a, the magnitude depending upon $K(x)$. Taking the Kac case, $K(x) = e^{-x^2}$, we obtain

$$\text{(3.30.10)} \qquad 2 \int_0^{a/2} e^{-x^2} dx \geq \lambda_M \geq 2 \int_0^a e^{-x^2} dx - \frac{1}{2} + \frac{e^{-a^2}}{a}$$

which yields the results

$$.843 \geq \lambda_M (2)/\pi^{1/2} \geq .713,$$

$$.995 \geq \lambda_M (4)/\pi^{1/2} \geq .749,$$

$$.999 \geq \lambda_M (10)/\pi^{1/2} \geq .899.$$

Notice that even for small a, (3.30.10) yields a rough idea of the true value of λ_M.

3.31 Second Proof

The method we present below yields the following useful result:

THEOREM 2. If $K(x)$ is non-negative, continuous, even, and

monotone decreasing for $0 < x < \infty$, the characteristic function $f_M(x)$ associated with λ_M, which we normalize by the requirement that $\int_0^a f_M(x)dx = 1$, possesses the following properties:

$$\text{(3.31.1)} \quad \text{(a) } f_M(x) = f_M(a - x) \,,$$

$$\text{(b) } f_M \text{ is monotone increasing in } 0 \le x \le a/2.$$

PROOF. We require the following two lemmas, the first of which is a well-known result in the theory of integral equations:

LEMMA 3. Let $K(x,y)$ be a continuous symmetric function defined over the square $0 \le x, y \le a$, and $g(x)$ be continuous over $0 \le x \le a$. Then, if we define

$$\text{(3.31.2)} \quad Tg = \int_0^a K(x,y)g(y)dy \,,$$

the limit

$$\text{(3.31.3)} \quad \lim_{n \to \infty} \frac{T^n g}{\lambda^n_M} = \emptyset(x)$$

exists and is a characteristic function of $K(x,y)$ associated with (3.31.2) provided that it is not identically zero.

LEMMA 4. If $f(x)$ has the following properties:

$$\text{(3.31.4)} \quad \begin{aligned} &\text{(a) } f(x) = f(a - x) \,, \\ &\text{(b) } f'(x) \ge 0 \text{ for } 0 \le x \le a/2 \,, \\ &\text{(c) } f(0) \ge 0 \,, \end{aligned}$$

then

$$\text{(3.31.5)} \quad Tf = \int_0^a K(x - y)f(y)dy$$

possesses the same properties, provided that $K(x)$ is even, non-negative monotone decreasing in the interval $(0,a)$, and possesses a derivative in this interval.

PROOF OF LEMMA 4. We have

$$g(x) = Tf = 2 \int_0^{a/2} (K(x - y) + K(a - x - y)f(y)dy \tag{3.31.6}$$

whence

$$g'(x) = 2 \int_0^{a/2} (K'(x - y) - K'(a - x - y))f(y)dy \quad . \tag{3.31.7}$$

Integration by parts yields

$$\begin{aligned} g'(x) &= 2f(0)(K(x) - K(a - x)) \\ &\quad + 2 \int_0^{a/2} (K(x - y) - K(a - x - y)f'(y)dy \quad . \end{aligned} \tag{3.31.8}$$

If $0 \le x, y \le a/2$, we have $x \le a - x$, $|x - y| \le a - x - y$, and consequently $K(x) \ge K(a - x)$, $K(x - y) \ge K(a - x - y)$. Therefore $g'(x) \ge 0$, with equality at $x = a/2$.

We now combine Lemmas 3 and 4 to prove Theorem 2. Let $f_0 = 1$, and define

$$f_{n+1} = \int_0^a K(x - y)f_n(y)dy \quad . \tag{3.31.9}$$

From Lemma 2 it follows that each $f_n(x)$ possesses properties Eqs. 3.31.4 (a), (b), and (c), since f_0 does trivially. Lemma 3 tells us that

$$\phi(x) = \lim_{n \to \infty} f_n(x)/\lambda_M^n \tag{3.31.10}$$

is a characteristic function of $K(x-y)$ associated with λ_M, provided that it is not identically zero. That it is nontrivial follows from the fact that 1 as a positive function cannot be orthogonal to $f_M(x)$ which is also positive. It follows then that $f_M(x)$ possesses the stated properties, since there is only one characteristic function associated with λ_M.

The monotonicity property of $f_M(x)$ will play an important role in our second approximation technique. We shall not obtain as close a bound as before, however. Let us normalize our solution, which we know to be positive by the requirement $\int_0^a f(x)dx = 1$. Integrating both sides of our integral equation between 0 and a we obtain

$$(3.31.11) \qquad \lambda_M = \int_0^a \left[\int_0^a K(x-y)dx \right] f(y)dy$$

$$= 2 \int_0^{a/2} \left[\int_0^y K(u)du + \int_0^{a-y} K(u)du \right] f(y)dy \quad .$$

From (3.31.11) we derive

$$(3.31.12) \qquad 2c = \lambda_M = 4c \int_0^{a/2} f(x)dx$$

$$- 2 \int_0^{a/2} \left[\int_0^y K(u)du + \int_0^{a-y} K(u)du \right] f(y)dy$$

$$= 2 \int_0^{a/2} \left[c - \int_0^y K(u)du + c \right.$$

$$\left. - \int_0^{a-y} K(u)du \right] f(y)dy$$

$$= 2 \int_0^{a/2} \int_y^\infty K(u)f(y)dudy$$

$$+ 2 \int_0^{a/2} \int_0^\infty K(u)f(y)dudy \geq 0 \quad .$$

Thus for Y in $(0, a/2)$,

$$|\lambda_M - 2c| = 2c - \lambda_M \le 2\int_0^Y \int_y^\infty K(u)f(y)dudy \tag{3.31.13}$$

$$+ 2\int_Y^{a/2} \int_y^\infty K(u)f(y)dudy$$

$$+ \int_0^{a/2} \int_{a/2}^\infty K(u)f(y)dudy$$

$$\le 2\int_0^Y \int_0^\infty K(u)f(y)dudy + 2\int_Y^{a/2}\int_Y^\infty K(u)f(y)dudy$$

$$+ 2\int_0^{a/2} f(y)dy \int_{a/2}^\infty K(u)du$$

$$\le 2c\int_0^Y f(y)dy + \int_0^{a/2}\int_Y^\infty K(u)f(y)dudy$$

$$+ \int_{a/2}^\infty K(u)du$$

$$= 2c\int_0^Y f(y)dy + \int_Y^\infty K(u)du + \int_{a/2}^\infty K(u)du \quad .$$

The original estimate involved $8c$ in place of $2c$ above.

It remains to choose Y advantageously and estimate $\int_0^Y f(y)dy$. We have for $0 \le y \le a/2$, using the monotonic character of $f(x)$,

$$\frac{1}{2}\int_0^{a/2} f(x)dx \ge \int_y^{a/2} f(x)dx \ge f(y)\left(\frac{a}{2} - y\right) \quad , \tag{3.31.14}$$

and hence $f(y) \leq 1/(a-2y)$. Therefore

$$(3.31.15) \qquad \int_0^Y f(y)dy \leq Y/(a - 2Y).$$

If $Y \to \infty$ in such a way that $Y/a \to 0$ as $a \to \infty$, we see that $\lambda_M \to 2c$. Choosing Y so that $2cY/(a-2Y) = \int_Y^\infty K(u)du$, we obtain a best possible error term from this procedure. For example, if $K(x) = e^{-x^2}$, we obtain as $a \to \infty$

$$(3.31.16) \qquad |\lambda_M - 2c| = O\left(\frac{(\log a)^{1/2}}{a}\right)$$

which is inferior to the result stated in Theorem 1.

3.32 An Approximation Method for Small a

Referring to (3.31.5), we see that it is possible to improve our estimates for λ_M by choosing, in place of $f = g = 1$, functions which more nearly represent $f_M(x)$. Since we know the general form of $f_M(x)$ from Theorem 2, it would seem that two classes of functions which might yield good results are

$$(3.32.1) \qquad f(x) = 1 + cx(a - x), \quad c \geq 0,$$

and

$$(3.32.2) \qquad \begin{aligned} f(x) &= 1, & 0 \leq x \leq b < a/2, \\ &= c, & b \leq x \leq a - b, \\ &= 1, & a - b \leq x \leq a, \quad c \geq 1. \end{aligned}$$

In each of these cases the numerical work connected with approximating to the largest characteristic root of the kernel $e^{-(x-y)^2}$ will not be overly complicated since all the integrals that occur may be evaluated in terms of known functions.

3.33 On the Non-Negativity of Green's Functions

Consider the inhomogeneous equation

$$(3.33.1) \qquad u'' + q(x)u = f(x) \,, \quad u(0) = u(1) = 0 \,,$$

whose solution can be written in the form

$$(3.33.2) \qquad u = \int_0^1 K(x,y)f(y)dy \quad .$$

We wish to examine the sign of the kernel $K(x,y)$, which we shall call the Green's function of the equation under a suitable assumption concerning $q(x)$.

This problem has been investigated by Aronszajn and Smith, using the theory of reproducing kernels and the result we shall obtain in a special case of a general result contained in their paper. Since, however, the method we shall use is so simple, we feel that it is worth noting. Similar results may be obtained for equations of the form

$$(3.33.3) \qquad u_{xx} + u_{yy} + u_{zz} + q(x,y,z)u = f(x,y,z) \,,$$

under corresponding assumptions, either by means of the method we present here, or as consequences of the general results of Aronszajn and Smith.

3.34 Statement of Results

The result we shall demonstrate is

Theorem. Let $q(x)$ satisfy the condition

$$(3.34.1) \qquad q(x) \leq \pi^2 - d \,, \quad d > 0 \,,$$

where π^2 appears as the smallest characteristic value of the Sturm-Liouville problem

(3.34.2) $$u'' + \lambda u = 0$$
$$u(0) = u(1) = 0 \quad .$$

Then

(3.34.3) $$K(x,y) \le 0$$

for $0 \le x, \; y \le 1$.

3.35 Discussion

The condition in (3.34.1) asserts the negative definite nature of the quadratic form

(3.35.1) $$\int_0^1 u(u'' + q(x)u)dx = \int_0^1 (q(x)u^2 - u'^2)dx \quad .$$

It will be clear from what follows that the truth of the theorem hinges upon this fact.

The corresponding result for the equation of (3.34.2) is

THEOREM. - Consider the Sturm-Liouville equation

(3.35.2) $$u_{xx} + u_{yy} + u_{zz} + u = 0 \, , \quad u(x,y,z) \in D \, ,$$
$$u = 0 \, , \quad (x,y,z) \in B \, ,$$

where B is the boundary of the finite domain D.

If $q(x,y,z) \le \lambda_1 - d$, $d > 0$, where λ_1 is the smallest characteristic value of the problem above, then the Green's function associated with the operator

(3.35.3) $$u_{xx} + u_{yy} + u_{zz} + q(x,y,z)u$$

is non-positive.

The proof follows the same lines as that given for the one-dimensional case below. It is clear that a variety of boundary conditions can be imposed.

3.36 Proof of Theorem

Consider the problem of minimizing the inhomogeneous quadratic form

$$(3.36.1) \qquad J(u) = \int_0^1 (u'^2 - q(x)u^2 + 2f(x)u)dx ,$$

under the assumption of (3.34.1) concerning $q(x)$, over all function $u(x)$ which satisfy the conditions $u(0) = u(1) = 0$, and for which the integrals exist.

The positive definite nature of the quadratic terms ensures the existence of a minimum which is determined by the Euler-Lagrange equation, which is precisely (3.34.2).

A necessary and sufficient condition that $K(x,y)$ be non-positive is that $u(x) \le 0$ for $0 \le x \le 1$ whenever $f(x) \ge 0$ for $0 \le x \le 1$.

Suppose that the minimizing function, $u(x)$, possessed an interval (a,b) within which it was positive.

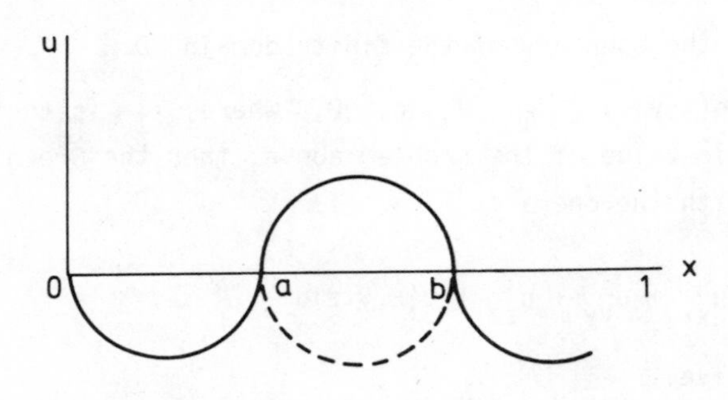

Fig. 3.1

Consider now the new function, shown in Fig. 3.1, obtained from $u(x)$ by retaining the values of $u(x)$ in the intervals where $u(x) \le 0$ and replacing $u(x)$ by $-u(x)$ in intervals where $u(x) \ge 0$. This does not change the value of the quadratic terms and decreases the integral $2\int_0^1 f(x)u\,dx$. Hence we have a contradiction to the statement that $u(x)$ yielded the minimum of $J(u)$. This change introduces discontinuities in the derivative $u'(x)$ which do not affect the integrability of $u'2$ and $q(x)u^2$.

3.37 On an Iterative Procedure for Obtaining the Perron Root of a Positive Matrix

The purpose of this section is to present an iterative procedure for obtaining the characteristic root of largest absolute value of a positive matrix.

The origin of the method is as follows. There is a result of von Neumann, a generalization of his fundamental min-max theorem in the theory of games, to the effect that

$$\text{(3.37.1)} \qquad \underset{y}{\text{Min}}\ \underset{x}{\text{Max}}\ \frac{(x,Ay)}{(x,By)} = \underset{x}{\text{Max}}\ \underset{y}{\text{Min}}\ \frac{(x,Ay)}{(x,By)}$$

where the variation is over the region defined by

$$\text{(3.37.2)} \quad R: \quad \begin{array}{ll} \text{(a)} & x_i \ge 0\ , \quad \sum_{i=1}^{n} x_i = 1\ , \\ \text{(b)} & y_i \ge 0\ , \quad \sum_{i=1}^{n} y_i = 1\ , \end{array}$$

and it is assumed that B has the property that

$$(x,\ By) \ge b \ge 0$$

for all $(x,y) \in R$.

It was observed by Shapley that this result can be obtained as a by-product of the theory of "games of survival", which requires only the fundamental min-max theorem, by considering the equation for λ,

$$(3.37.3)\qquad \begin{aligned}\lambda &= \min_y \max_x ((x,Ay) + \lambda(1 - (x,By)))\\ &= \max_x \min_y ((x,Ay) + \lambda(1 - (x,By))) ,\end{aligned}$$

where we impose the additional assumption that $1 > (x,By)$ for $(x,y) \in R$. This restriction is of no importance because of the homogeneity of the ratio in (3.37.1).

It is then easy to prove that there is a unique solution of (3.37.3) which may be obtained as the limit of the sequence $\{\lambda_n\}$ defined by

$$(3.37.4)\qquad \begin{aligned}\lambda_0 &= \min_y \max_x ((x,Ay)) = \max_x \min_y ((x,Ay))) ,\\ \lambda_{n+1} &= \min_y \max_x ((x,Ay) + \lambda_n (1 - (x,By)))\\ &= \max_x \min_y ((x,Ay) + \lambda_n (1 - (x,By))) ,\end{aligned}$$

and that this solution is then given also by the common value of the ratio in (3.37.1)

This procedure yields a theoretical and computational result which is quite useful. Furthermore, by means of this ingenious device we have a means of linearizing a number of problems relating to ratios. In this section we shall apply this idea to the problem of determining the root of a positive matrix of largest absolute value, using variational representation for this root involving ratio.

Let A be a square matrix (a_{ij}). It is called positive if $a_{ij} > 0$ for all i and j. The basic result concerning positive matrices is due to Perron and is the following:

LEMMA 1. If A is a positive matrix, there is a unique characteristic root of largest absolute value. This root is positive and its associated characteristic vector may be taken to be positive.

NOTATION. We denote this root by $p(A)$, the Perron root of A.

An alternative definition of this root, possessing the great merit of involving a variation is

LEMMA 2.

$$\begin{aligned} p(A) &= \operatorname*{Max}_{x} \operatorname*{Min}_{i} \sum_{j=1}^{n} a_{ij} x_j/x_i \\ &= \operatorname*{Min}_{x} \operatorname*{Max}_{i} \sum_{j=1}^{n} a_{ij} x_j/x_i \quad . \end{aligned} \tag{3.37.5}$$

This result has been used by several authors independently, and does not seem to have any particular known origin.

Here the variation is over the region defined by

$$x_i \geq 0 , \qquad \sum_i x_i = 1 \quad . \tag{3.37.6}$$

Let us show that Lemma 2 may be replaced by the stronger result

LEMMA 3. We have

$$\begin{aligned} p(A) &= \operatorname*{Max}_{R'} \operatorname*{Min}_{i} \sum_{j=i}^{n} a_{ij} x_j/x_i \\ &= \operatorname*{Min}_{R'} \operatorname*{Max}_{i} \sum_{j=1}^{n} a_{ij} x_j/x_i , \end{aligned} \tag{3.37.7}$$

where R' is defined by

$$x_i \geq d , \qquad \sum_i x_i = 1 , \tag{3.37.8}$$

and d is some parameter depending only upon A, Specifically, we may take

$$(3.37.9) \qquad d = \operatorname*{Min}_{i,j} a_{ij} \Big/ \operatorname*{Max}_{i} \left(\sum_{j=1}^{n} a_{ij} \right)$$

PROOF. The minimizing x_i constitute the characteristic vector associated with $p(A)$, normalized by the condition that $\sum_i x_i = 1$. Hence

$$(3.37.10) \qquad p(A)\, x_i = \sum_{j=1}^{n} a_{ij}\, x_j \,, \qquad i = 1,2,\ldots,n \quad .$$

Thus

$$(3.37.11) \qquad p(A) \operatorname*{Min}_{i} x_i \geq (\operatorname*{Min}_{i,j} a_{ij}) \sum_{j=1}^{n} x_j = \operatorname*{Min}_{i,j} a_{ij} \quad .$$

On the other hand,

$$(3.37.12) \qquad p(A) \operatorname*{Max}_{i} x_i \leq \operatorname*{Max}_{i} x_i \left(\operatorname*{Max}_{i} \sum_{j=1}^{n} a_{ij} \right) \quad ,$$

whence

$$(3.37.13) \qquad p(A) \leq \operatorname*{Max}_{i} \left(\sum_{j=1}^{n} a_{ij} \right) \quad .$$

Combining (3.37.11) and (3.37.13) we have

$$(3.37.14) \qquad \operatorname*{Min}_{i} x_i \geq \operatorname*{Min}_{i,j} a_{ij} \Big/ \operatorname*{Max}_{i} \left(\sum_{j=1}^{n} a_{ij} \right) \quad .$$

Let us now show, following the lead of Shapley, that we may define $p(A)$ as follows

LEMMA 4. $p(A)$ is the unique solution of

$$(3.37.15) \qquad \lambda = \operatorname*{Max}_{R'} \operatorname*{Min}_{i} \left[\sum_{j=1}^{n} a_{ij}\, x_j + \lambda(1 - x_i) \right]$$

or of

$$(3.37.16)\qquad \lambda = \underset{R'}{\text{Min}}\ \underset{i}{\text{Max}} \left[\sum_{j=1}^{n} a_{ij}\ x_j + \lambda(1 - x_i) \right]$$

where R' is as defined by (3.37.8).

PROOF. It is sufficient to prove that $p(A)$ satisfies (3.37.15). The proof of the other statement is similar. We have, for all x in R',

$$(3.37.17)\qquad \lambda \geq \underset{i}{\text{Min}} \left[\sum_{j=1}^{n} a_{ij}\ x_j + \lambda(1 - x_i) \right] ,$$

for any solution λ, with equality for at least one x. We shall prove below that there is exactly one solution which may be obtained iteratively.

Hence, for all $x \in R'$,

$$(3.37.18)\qquad 0 \geq \underset{i}{\text{Min}} \left[\sum_{j=1}^{n} a_{ij}\ x_j - \lambda\ x_i \right] ,$$

or

$$(3.37.19)\qquad 0 \geq \underset{i}{\text{Min}} \left[x_i \left\{ \sum_{j=1}^{n} a_{ij}\ x_j/x_i - \lambda \right\} \right] ,$$

for all $x \in R'$. Since $x_i > 0$, it follows that

$$(3.37.20)\qquad \lambda \geq \underset{i}{\text{Min}} \left(\sum_{j=1}^{n} a_{ij}\ x_j/x_i \right)$$

for all x, with equality for one x, at least. Hence

$$(3.37.21)\qquad \lambda = \underset{R'}{\text{Max}}\ \underset{i}{\text{Min}} \left(\sum_{j=1}^{n} a_{ij}\ x_j/x_i \right) = p(A) ,$$

which shows uniqueness provided we assume existence.

Similarly we may demonstrate the result in (3.37.16).

Let us now consider the nonlinear recurrence relation

$$u_{n+1} = \underset{R'}{\text{Min}} \ \underset{i}{\text{Max}} \left[\sum_{j=1}^{n} a_{ij} x_j + u_n (1 - x_i) \right] , \tag{3.37.22}$$

where R' is as above and u_0 is arbitrary. We shall prove

THEOREM.

$$p(A) = \lim_{n \to \infty} u_n \quad . \tag{3.37.23}$$

A similar result holds for the recurrence relation based upon (3.37.16).

PROOF. We have

$$\begin{aligned} u_{n+1} &= \underset{R'}{\text{Min}} \ \underset{i}{\text{Max}} \left[\sum_{j=1}^{n} a_{ij} x_j + u_n (1 - x_i) \right] \\ &= \underset{R'}{\text{Min}} \ \underset{y}{\text{Max}} \left[\sum_{i=1}^{n} y_i \left[\sum_{j=1}^{n} a_{ij} x_j + u_n \left(1 - x_i\right) \right] \right] \end{aligned} \tag{3.37.24}$$

where the maximum in y is over the region $y_i \geq 0$, $\sum_{i=1}^{n} y_i = 1$.

Using the min-max theorem of von Neumann, this may also be written

$$u_{n+1} = \underset{y}{\text{Max}} \ \underset{R'}{\text{Min}} \left\{ \sum_{i=1}^{n} y_i \left[\sum_{j=1}^{n} a_{ij} x_j + u_n \left(1 - x_i\right) \right] \right\} . \tag{3.37.25}$$

Let us write this recurrence relation in the form

$$u_{n+1} = \underset{R'}{\text{Min}} \ \underset{y}{\text{Max}} \ T(u_n, x, y) = \underset{y}{\text{Max}} \ \underset{R'}{\text{Min}} \ T(u_n, x, y) \quad . \tag{3.37.26}$$

Then using a device we have employed frequently in the theory of dynamic programming, we have

$$u_{n+1} = T(u_n, \bar{x}, \bar{y})$$

(3.37.27)

$$u_n = T(u_{n-1}, x^*, y^*)$$

where $(\bar{x}, \bar{y})$ and (x^*, y^*) are respectively values where the min-max and max-min are assumed for n and $n-1$ respectively.

Hence, by virtue of (3.37.26),

$$u_{n+1} = T(u_n, \bar{x}, \bar{y}) \geq T(u_n, \bar{x}, y^*)$$

(3.37.28)

$$\leq T(u_n, x^*, \bar{y}) ,$$

and

$$u_n = T(u_{n-1}, x^*, y^*) \geq T(u_{n-1}, x^*, \bar{y})$$

(3.37.29)

$$\leq T(u_{n-1}, \bar{x}, y^*) .$$

From this we obtain

$$u_{n+1} - u_n \geq T(u_n, \bar{x}, y^*) - T(u_{n-1}, \bar{x}, y^*)$$

(3.37.30)

$$\leq T(u_n, x^*, \bar{y}) - T(u_{n-1}, x^*, \bar{y}) ,$$

which yields

$$u_{n+1} - u_n \geq (u_n - u_{n-1}) \sum_{i=1}^{n} y_i^* (1 - \bar{x}_i)$$

(3.37.31)

$$\leq (u_n - u_{n-1}) \sum_{i=1}^{n} \bar{y}_i (1 - x_i^*) .$$

Hence

$$|u_{n+1} - u_n|$$

(3.37.32)
$$\leq |u_n - u_{n-1}| \text{ Max} \left[\sum_{i=1}^{n} y_i^*(1-\bar{x}_i) \sum_{i=1}^{n} \bar{y}_i(1-x_i^*) \right] .$$

Since $\bar{x}_i$, $x_i^* \geq d \geq 0$, and $y_i \geq 0$, $\sum_i y_i = 1$, we have

(3.37.33) $$|u_{n+1} - u_n| \leq (1 - d) |u_n - u_{n-1}| ,$$

and hence geometric convergence of $\sum_{n=0}^{\infty} (u_{n+1} - u_n)$. The limit of u_n exists, and must equal $p(A)$. Observe that this is a situation where only the value of a game is of interest, if we wish only to determine $p(A)$. Consequently, an iterative procedure may be of some merit here.

If we set

(3.37.34) $$u_0 = \underset{R'}{\text{Min}} \underset{i}{\text{Max}} \sum_j a_{ij} x_j ,$$

we see that $u_1 \geq u_0$ and hence $u_{n+1} \geq u_n$, which ensures monotone convergence.

Similarly, if we have

(3.37.35) $$a_1 \leq a_{ij} \leq a_2 ,$$

and use the Perron roots of the associated matrices as initial approximations, we obtain monotone increasing and monotone decreasing sequences respectively.

In many cases, we have a non-negative matrix rather than a positive matrix. If the matrix describes a single system, we know that some power of the matrix will be positive.

3.38 On the Second Greatest Characteristic Root Of A Positive Definite Matrix.

Above, we considered the problem of obtaining upper and lower bounds for the largest characteristic root of a positive definite matrix. In the following sections, we show that similar methods may be used for the second greatest characteristic root.

In Sec. 3.39, we give sequences which converge to the second greatest characteristic root from above and below. In Sec. 40, we present another sequence which may be used. In Sec. 41, we show how the modified Kronecker sum and product may be used. Finally, in Sec. 42, we show how similar methods may be used for the second smallest characteristic root of a second order differential equation.

3.39 Determination Of $\lambda 2$:

Let us use the notation of (3.37.1) and (3.37.2). We let A denote the positive definite matrix, and $\lambda_1 > \lambda_2 > \ldots$ denote the characteristic roots. Let $U_n = tr(A^n)$. Let us assume that the methods given in (3.37.1) and (3.37.2) determine λ_1 very accurately. Let $V_n = U_n - \lambda_1^n$ (1). Then, we may use V_n as we did in (3.37.1) and (3.37.2) to determine λ_2.

3.40 Determination Of $\lambda_1\lambda_2$:

Consider the sequence

$$W_n = \frac{U_n^2 - U_{2n}}{2} = (\lambda_1\lambda_2)^n + \ldots\ldots$$

we may use W_n as in (3.37.1) and (3.37.2) to determine $\lambda_1\lambda_2$. If, as we have assumed, λ_1 has been determined accurately, this gives an accurate determination of λ_2.

3.41 The Modified Kronecker Sum And Product:

The modified Kronecker product yields $\lambda_i \lambda_j$, $i \neq j$. The modified Kronecker sum yields $\lambda_i + \lambda_j$, $i \neq j$. In this way we can obtain estimates for λ_2, assuming that we know λ_1. The dimension of the modified Kronecker product and sum is quite high but the structure of the components is quite simple.

3.42 Characteristic Values:

In Sec. 3.11, we estimate for the smallest characteristic root of a second order linear differential equation. Methods similar to those used in Secs. 3.39 and 3.40 above may be used to determine the second smallest characteristic root.

Bibliography and Comments

Section 3.1. The result we use concerning matrices may be found in Bellman, R., Introduction to Matrix Analysis, McGraw-Hill Book Company, New York, 1970.

The results we use concerning inequalities may be found in Beckenbach, E. F., and Bellman, R., Inequalities, Springer-Verlag, Berlin, 1970.

Section 3.9. The background for Sturm-Liouville equations may be found in the following:

Courant, C., and Hilbert, D., Methods of Mathematical Physics, Interscience Publications, New York, 1953.

Section 3.12. The results given here were first presented in Bellman, R., "On the Determination of Characteristic Values for a Class of Sturm-Liouville Problems, "Illinois Journal of Mathematics, Vol. 2, 1958, pp. 577-585.

The results concerning entire functions may be found in Titchmarsh, E. C., Theory of Functions, Oxford University Press, England, 1939.

Section 3.23. The results given here were first presented in Bellman, R., "The Rayleigh Quotient and Dynamic Programming, International Journal of Mathematics and Physics, Vol. 1, No. 4, 1978.

The results concerning dynamic programming we used may be found in Bellman, R., Introduction to the Mathematical Theory of Control Processes, Vol. II, Academic Press, Inc., New York, 1971.

Dreyfus, S., and Law, A. M., The Art and Theory of Dynamic Programming, Academic Press, Inc., New York, 1977.

Section 3.30. The results here were first given in Bellman, R., and Latter, R., "On the Integral Equation $\lambda f(x) = \int_0^a K(x-y)f(y)dy$, Proceedings of the American Mathematical Society, Vol. 3, No. 6, December 1952, pp. 884-891.

Another application of the theorem of Jentsch may be found in the paper, Bellman, R., Soong, T. T., and Vasudevan, R., "On Moment Behavior of a Class of Stochastic Equations, "Journal of Mathematical Analysis and Applications, Vol. 40, No. 2, 1972, pp. 286-299.

A proof of the matrix version, due to Bohnenblust is given in the book on matrix analysis cited above.

Section 3.34. The result was first given in Bellman, R., "On the Non-Negativity of Green's Functions, "Boll. D'Unione Matematico, Vol. 12, 1957, pp. 411-413.

The result for quadratic variational problems may be found in Bellman, R., Introduction to the Mathematical Theory of Control Processes, Vol. I, Academic Press, Inc., New York, 1968.

Section 3.38. The results were first given in Bellman, R., "On an Iterative Procedure for Obtaining the Perron Root of a Positive Matrix, "Proceedings of the American Mathematical Society, Vol. 6, 1955, pp. 719-725.

Chapter IV

ON THE DETERMINATION OF CHARACTERISTIC VALUES FOR A CLASS OF STURM-LIOUVILLE PROBLEMS

4.1 Introduction

In this chapter we are interested in the problem of determining the characteristic values of the Sturm-Liouville equation,

$$u'' + \lambda a(t)u = 0 , \quad u(0) = u(1) = 0 . \tag{4.1.1}$$

It will be clear from what follows that the methods we discuss can be applied to equations of this type involving quite general boundary conditions, as long as the interval is finite.

There are, at present, a number of powerful techniques available for treating problems of this genre, based upon variational techniques, and upon matrix techniques applied to a finite difference version of the foregoing differential equation.

The variational approach depends upon the fact that if $a(t)$ satisfies a reasonable condition such as

$$0 < a^2 < a(t) < b^2 , \quad 0 < t < 1 , \tag{4.1.2}$$

then the characteristic values, $\lambda_1 < \lambda_2 < . . . ,$ are the respective relative minima of the functional

$$J(u) = \int_0^1 u'^2 \, dt / \int_0^1 a(t)u^2 dt \tag{4.1.3}$$

as u ranges over the space of functions for which the integrals exist and for which $u(0) = u(1) = 0$.

In particular,

$$\lambda_1 \le \int_0^1 u'^2 \, dt \Bigg/ \int_0^1 a(t)u^2 \, dt \tag{4.1.4}$$

for all functions $u(t)$ satisfying the prescribed boundary conditions. We thus have a means of obtaining upper bounds for λ_1 which turn out to be remarkably accurate even for simple choices of trial functions $u(t)$.

Another method is based upon using equations of the form

$$u_{n+2} - 2u_{n+1} + u_n + \lambda\Delta^2 a_n u_n = 0 , \tag{4.1.5}$$

$u(0) = u(N) = 0$, and applying any of a number of methods used to derive the characteristic roots and vectors of a symmetric matrix. For a detailed discussion of these methods, and others, we refer to the book by Collatz.

There is, however, a significant difference between a problem of this type, and the Sturm-Liouville problem described above. This is due to the fact that it is quite easy to find asymptotic solutions to (4.1.1) for large λ, and thus, approximate expressions for the higher characteristic values.

Let, for simplicity of notation, $a(t) = q^2(t)$; then the Liouville transformation $s = \int_0^t q(t_1)dt_1$, converts

$$u'' + \lambda q^2(t)u = 0 \tag{4.1.6}$$

into

$$\frac{d^2u}{ds^2} + \frac{q'(t)}{q^2(t)} \frac{du}{ds} + \lambda u = 0 \quad . \tag{4.1.7}$$

The further transformation

$$(4.1.8) \qquad v = u\sqrt{q(t)} = u \exp\left\{1/2 \int \frac{q'(t)}{q^2(t)}\, ds\right\}$$

converts (4.1.7) into

$$(4.1.9) \qquad \frac{d^2v}{ds^2} + \left[\lambda - 1/2\,\frac{d}{ds}\left(\frac{a'(t)}{a^2(t)}\right) - 1/4\left(\frac{a'(t)}{a^2(t)}\right)^2\right] v = 0 \quad .$$

The new boundary conditions are

$$(4.1.10) \qquad v(0) = 0 \; , \qquad v\left(\int_0^1 q(t)\, dt\right) = 0 \quad .$$

Writing (4.1.9) in the form

$$(4.1.11) \qquad v''(s) + (\lambda + b(s))\, v(s) = 0 \; ,$$

we know that we can find asymptotic developments for $v(s)$ starting from the integral equation

$$(4.1.12) \qquad v(s) = c_1 \cos \lambda^{1/2} s + c_2 \sin \lambda^{1/2} s - \int_0^s \frac{\left[\sin\lambda^{1/2}(s-r)\right]}{\lambda^{1/2}}\, b(r)v(r)\, dr$$

and iterating. Approximate values of λ are now determined by means of the constraint $v(\int_0^1 q(t)\, dt) = 0$. Thus, the higher characteristic values have the principal term

$$(4.1.13) \qquad \lambda_n \simeq n^2\, \pi^2 \Big/ \left(\int_0^1 q(t)\, dt\right)^2 \quad .$$

To obtain more precise results, we can use further terms of the asymptotic series derived from (4.1.12), and we can combine this with numerical integration of (4.1.1).

It follows from these considerations that the greatest difficulty is experienced in obtaining accurate estimations of the first characteristic value. In many investigations this is all that is desired.

We wish to present a new method, suitable for hand or digital computer calculation, which furnishes monotone convergence, through sequences of upper and lower bounds, to the smallest characteristic value. Similar sequences can be used to obtain monotone convergence to products of the form $\prod_{i=1}^{k} \lambda_i$. The method has the advantage of permitting λ_1 to be determined to a high degree of accuracy.

To illustrate these techniques, we consider the equation

$$(4.1.14) \qquad u'' + \lambda(1 + t)u = 0 , \qquad u(0) = u(1) = 0$$

which is connected with Airy's function, or Bessel functions of order 1/3.

4.2 The Equation Determining The Characteristic Values

Let us note in passing that the method we used is an application of an approach we have used, in various lecture courses on differential equations, to derive the fundamental results of Sturm-Liouville theory.

Consider the linear differential equation

$$(4.2.1) \qquad u'' + \lambda a(t)u = 0 , \qquad u(0) = 0 , \qquad u'(0) = 1 \quad .$$

The solution of this initial value problem may be obtained over $0 < t < 1$ as a power series in λ in the form

$$(4.2.2) \qquad u = t + \sum_{n=1}^{\infty} u_n(t) \lambda^n ,$$

where the sequence of coefficient functions $\{u_n(t)\}$, $n = 1,2,. . .$ may be determined by means of the recurrence relations

$$(4.2.3) \qquad u_0(t) = t ,$$

$$u_n(t) = - \int_0^t (t - s)u_{n-1} (s)a(s)ds , \quad n = 1,2,. . .$$

It is easy to see that u, as defined by (4.2.2), is an analytic function of λ for all finite λ for $0 < t < 1$. The roots of the equation

$$f(\lambda) = u(1) = 1 + \sum_{n=1}^{\infty} u_n(1)\,\lambda^n = 0 \tag{4.2.4}$$

are the desired characteristic values.

4.3 Discussion

If we assume that the sequence of coefficients is determined by means of either a hand or machine computation, a matter we will discuss again below, there is the problem of determining the first few roots of the equation in (4.2.4).

This is a problem which can be treated in several ways. It would seem that an efficient procedure would be to use the sequences we shall describe presently to obtain reasonably accurate estimates for the characteristic values, and then use the Newton-Raphson's method, or a modification, to obtain very accurate values.

4.4 Analytic Preliminaries

Referring to the equation in (4.2.1), we easily see that

$$|u(t)| \le e^{k|\lambda|^{1/2}} \quad . \tag{4.4.1}$$

for $0 < t < 1$, where k is a constant. Consequently, the Weierstrass factorization of $f(\lambda)$ takes the form

$$f(\lambda) = \prod_{i=1}^{\infty} (1 - \lambda/\lambda_i) \quad . \tag{4.4.2}$$

As we know, $\lambda_n = O(n^2)$ as $n \to \infty$, in view of the assumptions we have made concerning $a(t)$ in (4.1.2).

Our aim is now, by following the technique used by Newton to relate the sums of the powers of the roots and the elementary symmetric functions, which are the coefficients, to obtain relations for the sums

$$(4.4.3) \qquad b_r = \sum_{i=1}^{\infty} 1/\lambda i^r , \qquad r = 1,2, \ldots$$

in terms of the coefficients $u_n(1)$.

It is clear that

$$(4.4.4) \qquad \log f(\lambda) = \sum_{i=1}^{\infty} \log (1 - \lambda/\lambda_i) = - \sum_{r=1}^{\infty} (\lambda^r/r) \{ \sum_{i=1}^{\infty} 1/\lambda_i \}$$

$$= - \sum_{r=1}^{\infty} \lambda^r b_r/r ,$$

for $|\lambda| < \lambda_1$.

It is important then to obtain the coefficients of the expression of $\log f(\lambda)$. Although this can be done directly, it is easier to proceed as follows. Write

$$(4.4.5) \qquad \log f(\lambda) = \sum_{k=1}^{\infty} c_k \lambda^k .$$

Then

$$(4.4.6) \qquad f'(\lambda)/f(\lambda) = \sum_{k=1}^{\infty} kc_k \lambda^{k-1} ,$$

whence

$$(4.4.7) \qquad \sum_{n=1}^{\infty} nu_n(1) \lambda^{n-1} = \left(\sum_{k=1}^{\infty} kc_k \lambda^{k-1} \right) \left(1 + \sum_{n=1}^{\infty} u_n(1) \lambda^n \right),$$

whence we obtain the well-known recurrence relations

$$(4.4.8) \qquad nu_n = nc_n + \sum_{k=1}^{n-1} kc_k u_{n-k} .$$

These permits us to calculate the c_n in a very simple fashion once the sequence $\{u_n(1)\}$ has been determined, and thence the b_n.

4.5 Inequalities

Let us now show that the sequence $\{b_k\}$ can be used to obtain sequences which converge monotonically from above and below to the first characteristic value λ_1.

THEOREM 1. We have the inequalities

$$(4.5.1) \qquad b_k/b_{k+1} > \lambda_1 > 1/b_k^{1/k}, \qquad k = 1,2, \ldots .$$

The sequence $\{b_k/b_{k+1}\}$ is monotone decreasing; the sequence $\{1/b_k^{1/k}\}$ is monotone increasing, and

$$(4.5.2) \qquad \lambda_1 = \lim_{k \to \infty} b_k/b_{k+1} = \lim_{k \to \infty} 1/b_k^{1/k} \quad .$$

PROOF. The monotone character of the ratio b_k/b_{k+1} follows directly from Schwarz's inequality, since

$$(4.5.3) \qquad b_k^2 = \left(\sum_{i=1}^{\infty} 1/\lambda_i^k \right)^2 = \left(\sum_{i=1}^{\infty} 1/\lambda_i^{(k+1)/2} \lambda_i^{(k-1)/2} \right)^2$$

$$\leq \left(\sum_{i=1}^{\infty} 1/\lambda_i^{k+1} \right) \left(\sum_{i=1}^{\infty} 1/\lambda_i^{k-1} \right) = b_{k+1} b_{k-1} \quad .$$

The monotone behavior of $b_k^{1/k}$ is a consequence of the well-known inequality

$$(4.5.4) \qquad \left(\sum_{i=1}^{\infty} x_i \right) > \left(\sum_{i=1}^{\infty} x_i^2 \right)^{1/2} > \left(\sum_{i=1}^{\infty} x^3 , \right)^{1/3} > \ldots ,$$

for any set of non-negative x_i.

The proof of the limiting relation is clear.

4.6 Rate of Convergence

Since

$$(4.6.1) \qquad \frac{b_k}{b_{k+1}} = \frac{(1/\lambda_1^k)\ [1 + (\lambda_1/\lambda_2)^k + \ldots]}{(1/\lambda_1^{k+1})\ [1 + (\lambda_1/\lambda_2)^{k+1} + \ldots]}$$

$$= \lambda_1\ [1 + (\lambda_1/\lambda_2)^k - (\lambda_1/\lambda_2)^{k+1} + \ldots]\ ,$$

we see that

$$(4.6.2) \qquad b_k/b_{k+1} - \lambda_1 \simeq \lambda_1\ (\lambda_1/\lambda_2)^k$$

for large k.

Similarly,

$$(4.6.3) \qquad b_k^{1/k} = (1/\lambda_1)(1 + (\lambda_2/\lambda_2)^k + \ldots)^{1/k}$$

$$\simeq (1/\lambda_1)(1 + k(\lambda_1/\lambda_2)^k)$$

for large k.

It is to be expected that b_k/b_{k+1} will furnish a better approximation to λ_1 for large k.

4.7 Discussion

For the case where $a(t) \equiv 1$, $\lambda_1/\lambda_2 = 1/4$. Consequently, in general, the rate of convergence of these sequences will not be too rapid. There are two things we can do to obtain more accurate estimations of λ_1. In the first place, we can use the root-squaring technique. Since

$$(4.7.1) \qquad f(\lambda) = \prod_{i=1}^{\infty} (1 - \lambda/\lambda_i)\ ,$$

we see that

$$(4.7.2) \qquad f_1(\lambda) = f(\lambda^{1/2})f(-\lambda^{1/2}) = \prod_{i=1}^{\infty} (1 - \lambda/\lambda_i^2) \quad .$$

Using the power series development for $f_1(\lambda)$ we obtain a sequence b'_k with

$$(4.7.3) \qquad \lim_{k \to \infty} b'_k/b'_{k+1} = \lambda_1^2 ,$$

and a rate of convergence depending upon $(\lambda_1/\lambda_2)^2$.

Alternatively, once we have an estimate for λ_1 with an accuracy of 1 in 10^{-2}, we can then turn to the power series for $f(\lambda)$ and use the Newton approximation technique,

$$(4.7.4) \qquad \lambda_1^{(n+1)} = \lambda_i^{(n)} - f(\lambda_1^{(n)}) \Big/ f'(\lambda_1^{(n)}) \quad .$$

This will yield a further approximation with accuracy of essentially 1 in 10^{-24}. Continued use of this technique is limited only by the number of $u_n(1)$ which are computed, and the accuracy of this computation. There is no difficulty involved in using this technique here, since we know from theoretical consideration that the roots of $f(\lambda)$ are simple.

4.8 Inequalities for $\prod_{i=1}^{R+1} \lambda_i$

Similar upper bounds can be obtained for the products $\prod_{i=1}^{R+1} \lambda_i$, $R = 1,2, \ldots$.

Consider the determinant

$$(4.8.1) \qquad b_k^{(R)} = \begin{vmatrix} b_k & b_{k+1} & \cdots & b_{k+R} \\ . & . & \cdots & . \\ . & . & \cdots & . \\ . & . & \cdots & . \\ b_{k+R} & b_{k+R+1} & \cdots & b_{k+2R} \end{vmatrix} , \qquad R = 1,2,\ldots$$

It is not difficult to show that

$$(4.8.2)\qquad \lim_{k\to\infty} b_k^{(R)} \Big/ b_{k+1}^{(R)} = \lim_{k\to\infty} (b_k^{(R)})^{-1/k} = \lambda_1\lambda_2 \; . \; . \; . \; \lambda_{R+1} \; .$$

To show that

$$(4.8.3)\qquad b_k^{(R)} \Big/ b_{k+1}^{(R)} > b_{k+1}^{(R)} \Big/ b_{k+2}^{(R)} \; , \qquad k = 1,2, \; . \; . \; . \; ,$$

for $R = 1,2, \; . \; . \; . \; ,$ we use the well-known fact that the matrix

$$(4.8.4)\qquad B_k^{(R)} = \begin{vmatrix} b_k & b_{k+1} & \cdots & b_{k+R} \\ . & . & \cdots & . \\ . & . & \cdots & . \\ . & . & \cdots & . \\ b_{k+R} & b_{k+R+1} & \cdots & b_{k+2R} \end{vmatrix}$$

is positive definite for all k and R, and hence that $(B_k^{(R)})^{-1}$ is positive definite.

The sequence $(b_k^{(R)})^{-1/k}$ does not seem to have any simple monotonicity properties.

4.9 The Equation $u'' + \lambda(1 + t)u = 0$

Let us now illustrate some of the ideas discussed above by means of the equation

$$(4.9.1)\qquad u'' + \lambda(1 + t)u = 0 \; , \qquad u(0) = u(1) = 0 \quad .$$

The first problem we face is that of computing the sequence $\{u_n(t)\}$ by means of the recurrence relations of (4.2.3). Since $u(t)$ is an entire function of λ for $0 \leq t \leq 1$, the coefficients, $u_n(t)$, become quite small as n increases. If $a(t) \equiv 1$, the coefficient of λ^n is $(-1)^n/(2n+1)!$ Hence, if we are using a digital computer, even one with floating point arithmetic, it is necessary to renormalize. A very simple renormalization is one which sets

$$v_n(t) = (-1)^n(2n+1)!\, u_n(t) \quad . \tag{4.9.2}$$

Then

$$v_n(t) = \frac{1}{(2n+1)2n} \int_0^t (t-s)v_{n-1}(s)a(s)ds \ , \tag{4.9.3}$$

$$v_0(t) = 1 \quad . \qquad\qquad n = 1,2,\ldots,$$

Since (4.9.3) is equivalent to the differential recurrence relation

$$v_n''(t) = a(t)\, v_{n-1}(t) \Big/ 2n(2n+1) \qquad v_n(0) = v_n'(0) = 0 \ , \tag{4.9.4}$$

we can use a Runge-Kutta integration procedure to obtain fairly accurate values of $v_n(1)$ (see Table 4.1).

Table 4.1

n	$v_n(1) = (-1)^n(2n+1)!\, u_n(1)$
0	1. 000 000 000
1	1. 499 999 92
2	2. 238 094 66
3	3. 333 330 15
4	4. 960 358 93
5	7. 378 146 87
6	10. 971 261 4
7	10. 310 824 0
8	24. 244 529 3
9	36. 028 967 6
10	53. 522 379 4

The decision as to how many elements of the sequence $\{u_n(1)\}$ to compute depends upon an *a priori* estimate of the magnitude of λ_1, the time involved in the computation, the accuracy of the computation, and the accuracy with which λ_1 is desired.

Since $1 + t \geq 1$, we see that $\lambda_1 < \pi^2 < 10$. Hence the order of magnitude of the last term computed in the power series would be

$$(4.9.5) \qquad u_n(1)\lambda_1^n < \frac{53.5}{(21)!} 10^{10} < \frac{10^2 \cdot 10^{10}}{(20)!} < \frac{10^2 \cdot 10^{10}}{2^{20} \cdot 10^{20}} = \frac{10^{-8}}{2^{20}}$$

(using Sterling's approximation). This is more than sufficient, considering the inaccuracy involved in numerical integration, for the determination of λ_1, and is sufficient for the determination $\lambda_2 \leq 4\pi^2$.

The next step is to compute the sequence of coefficients in $\log f(\lambda)$, namely $\{b_k\}$, using (4.4.8). The results are given in Table 4.2, together with the ratios b_k/b_{k+1} and the roots $b_k^{-1/k}$.

Table 4.2

k	b_k	$\frac{b_k}{b_{k-1}}$	$b_k^{-1/k}$ (slide rule evaluation)
1	25.0000	$9.921\ 26 \times 10^{-2}$	4.00×10^{-2}
2	251.984	$6.958\ 90 \times 10^{-2}$	6.30×10^{-2}
3	3621.03	$6.632\ 47 \times 10^{-2}$	6.51×10^{-2}
4	54595.5	$6.567\ 79 \times 10^{-2}$	6.54×10^{-2}
5	831261.0	$6.553\ 06 \times 10^{-2}$	6.55×10^{-2}
6	12685100.0	$6.549\ 54 \times 10^{-2}$	-
7	193679×10^3	$6.548\ 66 \times 10^{-2}$	-
8	29575×10^5	-	-

For the purposes of using the Newtonian scheme mentioned above (4.7.4), we see that b_4/b_5 and $b_4^{-1/4}$ yield sufficiently good initial approximations with an error of about 1 in 600. One or two applications of (4.7.4) would yield λ_1 to an accuracy sufficient for most purposes.

The convergence of the sequences for $\lambda_1\lambda_2$ is much less rapid as is to be expected. The results are shown in Table 4.3.

Table 4.3

k	$b_k^{(1)} = b_k\, b_{k+2} - b_{k+1}^2$;	$b_k^{(1)} / b_{k+1}^{(1)}$;	$(b_k^{(1)})^{-1/k}$
1	27030.0	418.85×10^{-4}	$.37 \times 10^{-4}$
2	645330.0	219.85×10^{-4}	12.45×10^{-4}
3	29353×10^3	188.93×10^{-4}	32.45×10^{-4}
4	15537×10^5	179.34×10^{-4}	50.37×10^{-4}
5	86634×10^6	-	-

Using the value of λ_1 obtained above, we obtain a first approximation of $\lambda_2 \simeq 27$. From the monotonicity of the ratios, we know that λ_2 is actually less than this. An application of Newton's approximation will yield a greatly improved result.

Note that λ_2 is sufficiently large so that the asymptotic techniques discussed in 4.1 can be used to provide an independent check of the accuracy of the first approximation to λ_2.

4.10 Alternate Computational Scheme For Polynomial Coefficients

In what has preceded, we have spoken in terms of numerical evaluation of the sequence $\{u_n(t)\}$. Although this procedure has the great advantage of straightforwardness and simplicity, via hand computation or digital computation, it suffers from the fact that errors of integration arise, and grow with each new member of the sequence.

Consequently, it is worth noting a special, but important, case in which we can avoid mechanical quadrature and carry out the entire operation by hand.

Suppose that $a(t)$ is a polynomial of the form

$$(4.10.1) \quad a(t) = a_0 + a_1 t + \ldots + a_k t^k \quad .$$

It will be clear then that the elements of the sequence $\{u_n(t)\}$ will also be polynomials. Furthermore, it is clear that $u_n(t)$ will have the form

$$(4.10.2) \quad u_n(t) = a_0 t^{2n+1} \Big/ (2n+1)! + \ldots + a_{kn} t^{2n+k} + \ldots$$

Using the recurrence relation of (4.2.3), we can then obtain linear recurrence relations for the sequence $\{a_{kn}\}$, k= 1,2, . . . ; n = 1,2, . . .

There are a number of renormalization questions concerned with the effective calculation of the sequence, and asymptotic relations which can be used to speed the computation. A discussion of these would take us too far afield.

4.11 Extension To Higher Order Equations

Let us now consider the equation

$$(4.11.1) \quad u^{(4)} + \lambda a(t) u = 0$$

with the boundary conditions

$$(4.11.2) \quad u(0) = u'(0) = 0 \ , \qquad u(1) = u'(1) = 0 \quad .$$

Proceeding as above, we consider the solution, $u(t, \lambda)$, of the initial value problem

$$(4.11.3) \qquad u(0) = 0 \ , \quad u'(0) = 0 \ , \quad u''(0) = c_1 \ , \quad u'''(0) = c_2 \ ,$$

which we can write in the form

$$(4.11.4) \qquad u = c_1 u_1(t, \lambda) + c_2 u_2(t, \lambda) \ ,$$

where u_1 and u_2 are determined by the initial conditions

$$(4.11.5) \qquad u_1(0) = 0 \ , \quad u_2(0) = 0 \ , \quad u'_1(0) = 0 \ , \quad u'_2(0) = 0 \ ,$$

$$u''_1(0) = 1 \ , \quad u''_2(0) = 0 \ , \quad u'''_3(0) = 0 \ , \quad u'''_3(0) = 1 \ .$$

As before, there is no difficulty in obtaining the power series development in terms of λ for the functions u_1 and u_2.

Applying the boundary conditions in (4.11.2), we obtain the simultaneous equations

$$(4.11.6) \qquad c_1 u_1(1, \lambda) + c_2 u_2(1, \lambda) = 0 \ , \quad c_1 u_1'(1, \lambda) + c_2 u_2'(1, \lambda) =$$

whence the determining equation for λ is

$$(4.11.7) \qquad f(\lambda) = \begin{vmatrix} u_1(1, \lambda) & u_2(1, \lambda) \\ u_1'(1, \lambda) & u_2'(1, \lambda) \end{vmatrix} = 0 \ .$$

From here on, the argumentation is as before.

Chapter V

LINEAR DIFFERENTIAL EQUATIONS WITH CONSTANT COEFFICIENTS

5.1 Introduction

In this chapter, we consider the matrix exponential. We present two methods for calculating it. By using this matrix, we can avoid long term integration.

We show that the matrix exponential can be used to determine the inverse of a stability matrix. Then, we present a method of Ostrowski for determining an inverse of a matrix with positive characteristic roots. Finally, we show that some simple matrix relations enable us to extend many results to arbitrary matrices.

Consider the equation

$$\frac{dx}{dt} = Ax , \qquad x(0) = c \quad . \tag{5.1.1}$$

Here, A is a constant matrix.

The problem we wish to consider is that of determining x at a point without determining the intermediate values.

5.2 Standard Procedures

As is well-known, the solution of the equation above may be written

$$x = e^{At}c \quad . \tag{5.2.1}$$

If the dimension of A is small, the simplest procedure is to reduce the system to a single equation for one of the components. If the dimension of A is large, this procedure may be onerous.

We can also employ matrix theory to reduce A to a canonical form and thus solve the system. This procedure may be onerous because it involves finding the characteristic vectors of A.

We can also use straightforward integration routines. However, we do not want to integrate over a long interval because of possible round-off error. In addition, this is an unesthetic method because it generates a great deal of useless data.

5.3 Use of the Powers of Two

If x and A were scalar quantities, an ordinary function and an ordinary number, the solution of the first-order differential equation subject to an initial condition $x(0) = c$, could be written down immediately

$$x = e^{At}c \quad . \tag{5.3.1}$$

A representation of this nature holds in the multi-dimensional case where the matrix exponential e^{At} possesses analogues of some of the fundamental properties of the scalar exponential.

To begin with, the basic properties of linear differential systems ensures that we can write the solution of (5.3.1) subject to $x(0) = c$ in the form

$$x(t) = X(t)c \ , \tag{5.3.2}$$

where $X(t)$ is the matrix formed in the following fashion. Let $x^{(1)}$ denote the solution of (5.3.1) subject to

$$(5.3.3) \qquad x^{(1)}(0) = \begin{bmatrix} 1 \\ 0 \\ \cdot \\ \cdot \\ \cdot \\ 0 \end{bmatrix},$$

Let $x^{(2)}$ be the solution subject to

$$(5.3.4) \qquad x^{(2)}(0) = \begin{bmatrix} 0 \\ 1 \\ \cdot \\ \cdot \\ \cdot \\ 0 \end{bmatrix},$$

and so on, with $x^{(N)}$ determined by

$$(5.3.5) \qquad x^{(N)}(0) = \begin{bmatrix} 0 \\ 0 \\ \cdot \\ \cdot \\ \cdot \\ 1 \end{bmatrix}.$$

Let $X(t)$ now be the matrix whose N columns are $x^{(1)}$, $x^{(2)}, \ldots, x^{(N)}$,

$$(5.3.6) \qquad X(t) = \begin{pmatrix} x^{(1)} & x^{(2)} & \ldots & x^{(N)} \\ & & & \\ & & & \\ & & & \end{pmatrix}.$$

It follows that $X(t)$ solves the differential equation

$$X' = AX \,, \qquad X(0) = I \,, \tag{5.3.7}$$

where I is the identity matrix. We see then one reason for the notation we have adopted. It is, however, not the principal reason.

Let us now establish that

$$e^{A(t+s)} = e^{At} e^{As} \tag{5.3.8}$$

for $-\infty < s, t < \infty$. Perhaps the easiest, and certainly the most elegant, is to note that both are solutions of (5.3.7) satisfying the initial condition $X(0) = e^{As}$. Hence, by the basic uniqueness theorem, they are equal.

If we set $s = -t$, we have

$$I = e^{A(0)} = e^{At} e^{-At} \tag{5.3.9}$$

which establishes that e^{At} is never singular and that its inverse is e^{-At}, two more good reasons for the notation.

Let us note, however, that

$$e^{(A+B)t} \neq e^{At} e^{Bt} \tag{5.3.10}$$

in general does not hold. There is equality for all t if and only if A and B commute.

Using the basic functional equation for the exponential, we have

$$e^{At} = \left(e^{At/2^n} \right)^{2^n} \tag{5.3.11}$$

Here n is chosen so that the quantity $t/2^n$ is small.

We see that the determination of the exponential requires n matrix squarings, and the determination of $e^{At/2^n}$.

5.4 The Determination of e^{As}

To determine e^{As} for s small, we use the expansion

$$(5.4.1) \qquad e^{As} = I + As + A^2 s/2 + \ldots$$

We take as many terms of the expansion as desired for accuracy.

5.5 Another Method for Calculating the Matrix Exponential

The problem of stiffness can sometimes be circumvented by a direct calculation of the matrix exponential.

Let us note here another method of calculating the matrix exponential

$$(5.5.1) \qquad e^{A} = \lim_{n \to \infty} (I + \frac{A}{n})^n \quad .$$

If n is taken to be a power of two, the successive terms of the sequence can be calculated by squaring.

We thus have two different methods for calculating the matrix exponential. Since computers are so fast, we should routinely use several different methods for calculating everything. The results can then be compared. If the results are sufficiently close, within a specified bound, the calculation can proceed. If the results do not agree according to specification, we can give instructions that the computation is to stop.

It is also important theoretically to have as many different approaches as possible because different approaches generalize in different ways.

5.6 A Result Concerning Stability Matrices

Let A be a real matrix satisfying the condition

$$(5.6.1) \qquad a_{ij} \geq 0 , \qquad i \neq j \quad .$$

As is well known, this condition is a necessary and sufficient condition for the elements of the matrix exponential, e^{At}, to be non-negative.

From this it follows that the solution of

$$(5.6.2) \qquad \frac{dx}{dt} = Ax + f , \quad x(0) = c ,$$

is non-negative whenever the initial conditions and forcing function are. This is a fundamental result in economics and chemotherapy.

If we add the further condition that A is a stability matrix, we obtain a positivity result for the solution of an equation associated with the Lyapunov equation.

Finally, if we combine these two conditions on A, we obtain a nonpositivity result for the elements of the inverse of A.

5.7 The Lyapunov Equation

Let us now consider the equation

$$(5.7.1) \qquad \frac{dX}{dt} = AX + XB , \qquad X(0) = C \quad .$$

As is well known, the solution is given by

$$(5.7.2) \qquad X = e^{At} Ce^{Bt} \quad .$$

Let us now consider the inhomogeneous equation,

$$(5.7.3) \qquad \frac{dX}{dt} = AX + XB + F(t) , \qquad x(0) = C \quad .$$

The solution of this equation is given by

$$(5.7.4) \qquad X = e^{At} Ce^{Bt} + \int_0^t e^{A(t-s)} F(s) e^{B(t-s)} ds \quad .$$

If we impose (5.6.1), we see that the elements of X are non-negative if the initial matrix and the forcing matrix have non-negative elements.

Let us now consider the algebraic equation

$$AX + XB = C \quad . \tag{5.7.5}$$

We obtain the solution

$$X = -\int_0^\infty e^{At}\, C e^{Bt}\, dt \ , \tag{5.7.6}$$

if we assume that A and B are stability matrices. This solution is obtained by integrating the Lyapunov equation between 0 and ∞.

It may be shown by the use of Kronecker sums that the algebraic equation has a solution whenever the sum of two characteristic roots of A and B is not zero.

5.8 The Inverse Matrix

Let us assume, as above, that A satisfies the condition (5.6.1) and that it is a stability matrix. Then we have

$$-A^{-1} = \int_0^\infty e^{At}\, dt \quad . \tag{5.8.1}$$

The fact that A is a stability matrix means that the integral exists. A simple way to establish this relation is to start with

$$\frac{dX}{dt} = AX \ , \qquad X(0) = I \ , \tag{5.8.2}$$

and integrate between zero and ∞.

The condition (5.6.1) implies that the elements of the matrix exponential are non-negative. We see then that under these two hypotheses the elements of A^{-1} are nonpositive.

5.9 The Inverse of a Stability Matrix

Let us now only retain the hypothesis that A is a stability matrix. Then, we have (5.8.1)

The integral may be evaluated by quadrature. The identity gives us a way of finding a particular element in A^{-1}.

5.10 A Result of Ostrowski

Consider the identity

$$\frac{1}{1-x} = (1 + x)(1 + x^2)(1 + x^4) \ldots \tag{5.10.1}$$

valid for $|x| < 1$.

This identity may be readily established. It is an analytic equivalent of the fact that every positive integer is uniquely representable as a sum of distinct powers of two.

It was observed by Ostrowski that this relation holds if x is a matrix whose characteristic values are less than one in absolute value. If A is a matrix with positive characteristic roots, it may easily be modified so that it has the form

$$I - X \quad . \tag{5.10.2}$$

We thus have a way of obtaining the inverse of a matrix with positive characteristic value.

5.11 Inverses

We may use the foregoing procedure by virtue of the relation

$$A^{-1} = (A^2)^{-1} A \quad . \tag{5.11.1}$$

Similarly, we may use any technique for symmetric matrices by virtue of the relation

$$A^{-1} = (A'A)^{-1} A' \quad . \tag{5.11.2}$$

Bibliography and Comments

Section 5.1. For results we employ concerning matrix theory, see Bellman, R., *Introduction to Matrix Analysis*, McGraw-Hill Book Co., New York, 1970.

Section 5.7. This result was given in Bellman, R., "Some Consequences of the Non-negativity of the Elements of the Matrix Exponential", "*Journal on Nonlinear analysis: Theory, Methods and Appl.*, Vol. 4, No. 4, 1980, pp. 735-736.

Chapter VI

LINEAR DIFFERENTIAL EQUATIONS WITH VARIABLE COEFFICIENTS

6.1 Introduction

In this chapter, we discuss some questions where the coefficient matrix is variable.

First, we shall discuss the case where this matrix is periodic, giving the Floquent representation. This requires that we have a method for finding the exponential representation of a nonsingular matrix. We use a version of the Newton-Raphson technique.

If the coefficient matrix is doubly periodic, a classic result of Hermite tells us that all the solutions are doubly periodic. We show that this leads to a perturbation procedure.

The question of an asymptotic form of the solution is quite difficult. Different methods are available for different hypothesis for the coefficient matrix. We say a few words about this.

Then, we turn to the method of differential quadrature. A good deal of work is necessary here to make the method rigorous.

Next, we turn to the use of Laplace transform. If we combine this with a quadrature technique, we have a method for obtaining the solution of a linear differential equation using a system of approximate algebraic equations.

Finally, we consider some technique for finding a particular solution of a linear system.

6.2 Periodic Matrices

A very important case is that where the coefficient matrix is periodic. Let us assume that this period is one. In that case, we have a representation for the solution, the Floquent representation. Since it is so easy to establish, we shall give a proof here. Consider the equation

$$X' = A(t)x , \quad X(0) = I \quad . \tag{6.2.1}$$

We know the solution at (6.2.1) is nonsingular. Hence, it is an exponential e^B. Consider the expression $X(t)e^{-Bt}$. We see that it is periodic of period one. Hence, we have the Floquent representation

$$X = P(t)e^{Bt}. \tag{6.2.2}$$

Here $P(t)$ is periodic of period one.

If the dimension of the system is small, we can use a canonical representation of $X(1)$ to determine B. If the dimension is large, we can use the procedure given below. We shall also discuss the determination of $P(t)$.

6.3 A Nonsingular Matrix is an Exponential

It is well known that a nonsingular matrix is an exponential. This fact plays an important role in the Floquent representation where the coefficient matrix is periodic. However, the determination of the exponent is not simple.

Here, we shall use a matrix version of the Newton-Raphson formula for the one-dimensional case to find this. The relation we shall employ is

$$X_{n+1} = X_n + (I - Ae^{-X_n}) \quad . \tag{6.3.1}$$

Here, A is a positive definite matrix.

The key observation is that all the elements of the sequence commute with A. This may easily be proved inductively. Hence, they may all be reduced to diagonal form by means of the same transformation. The convergence is thus reduced to the one-dimensional case, where, as is well known, we have quadratic convergence.

Once we have established this relation for A, the calculation of the inverse is easily obtained. We have

$$A^{-1} = e^{-B} \quad . \tag{6.3.2}$$

We may use this result to solve the linear system

$$Ax = b \quad . \tag{6.3.3}$$

As we know there is no restriction in assuming A is positive definite. We may write the original system as

$$A'Ax = A'b \quad . \tag{6.3.4}$$

Here, a prime denotes the transpose of A.

6.4 The Determination of P(t)

The determination of $P(t)$ is not simple. We shall give three methods which may be viewed, with the usual recommendation that all three be used at the same time. In the first place, once we have determined B, we may use numerical integration to determine X, and thus $P(t)$.

In the second place, we may use the differential equation to obtain an infinite system of linear algebraic equations for the Fourier coefficients of $P(t)$. This infinite system must be solved by means of truncation. Since the Fourier coefficients decrease very rapidly, assuming that A has as many derivatives as desired, truncation should be very effective.

We may determine the Fourier coefficients directly, using a quadrature technique.

6.5 A Perturbation Procedure

Let us make some comments concerning linear equations with periodic coefficients. An equation such as

$$u'' + (a + b \cos t)u = 0 , \tag{6.5.1}$$

despite its simple appearance, cannot be solved explicitly in terms of the elementary functions. It defines a new class of transcendental functions, the Mathieu functions.

To treat equations of this nature various approximate techniques have been introduced. Let us discuss one that has been rather neglected. It consists of an initial complication that ultimately provides a simplification.

It was shown by Hermite as a corollary of a general result that the equation

$$u'' + (a + b \operatorname{cn} t)u = 0 , \tag{6.5.2}$$

where $\operatorname{cn} t$ is the Jacobian elliptic function that reduces to $\cos t$ as the modulus $k^2 \to 0$ could be solved explicitly in terms of doubly periodic functions, and thus in terms of the classical doubly periodic functions. The limit of this solution as $k^2 \to 0$ is the solution of (6.5.1).

Hence we have a systematic way of obtaining approximate solutions to linear equations with periodic coefficients by means of the exact solution of linear equations with doubly-periodic coefficients.

With the digital computer available, this is now a feasible procedure.

6.6 Asymptotic Behavior

In many cases, we are only interested in the asymptotic character of the solution.

If the system has dimension two, we can obtain a second-order linear differential equation for each component. We can now employ the Liouville transformation to put the equation into a convenient form which allows us to determine the asymptotic behavior. However, it is not a simple matter to determine the coefficients in this asymptotic form. We shall employ invariant imbedding in a later chapter to do this. If the dimension of the system is larger than two, we can employ the method of Levinson. Once again, however, it is not a simple matter to determine the coefficients in terms of the initial values. Although invariant imbedding can be used for this purpose, the necessary analytic work has not been carried out.

6.7 Differential Quadrature and Long-Term Integration

The term "quadrature", as ordinarily used, applies to the approximate evaluation of an integral

$$(6.7.1) \qquad \int_0^1 f(x)\,dx \cong \sum_{i=1}^{N} w_i f(x_i) \quad .$$

It has been shown that this technique can be utilized in a simple and systematic fashion to obtain the computational solution of nonlinear differential-integral equations derived from applications of the theory of invariant imbedding to transport processes. The foregoing approximation technique, however, can be extended to far more general linear functionals. Thus, we can write

$$(6.7.2) \qquad f'(x_i) \cong \sum_{j=1}^{N} a_{ij} f(x_j) , \qquad i = 1,2,\ldots,N \quad ,$$

with the coefficient matrix (a_{ij}) determined in various fashions. We call this procedure "differential quadrature". We indicated how this provided a new approach to the identification of parameters in systems described by various types of functional equations, a method quite different from the procedure based on the use of quasilinearisation.

In general, we can contemplate the systematic use of various approximation techniques to eliminate transcendental operations. This is part of a general theory of closure of operations, a theory which has become increasingly significant with the introduction of the hybrid computer. We wish to indicate how these ideas offer a new technique for the numerical solution of initial value problems for ordinary and partial differential equations with particular relevance to certain difficult questions arising from long-term integration.

6.8 Long-Term Integration

Consider the vector equation

$$y' = g(y) , \quad y(0) = c , \tag{6.8.1}$$

where y is an N-dimensional vector, and suppose that it is desired to calculate $y(T)$. A finite difference approximation (crude version),

$$w(t + \Delta) - w(t) = g(w(t))\Delta , \quad w(0) = c , \tag{6.8.2}$$

$t = 0, \Delta, 2\Delta, \ldots,$ leads to an algorithm well-suited to the nature of the contemporary digital computer. Starting with the initial value $w(0) = c$, we can use (6.8.2) in an iterative fashion to calculate in turn $w(\Delta), w(2\Delta), \ldots,$ and so on. We expect that $w(n\Delta) \cong y(n\Delta)$.

If T, the point in time at which the value of y is desired, is large, the foregoing procedure has several drawbacks. In the first place, we encounter the problem of numerical stability. An accumulation of evaluation and round-off errors may seriously contaminate $w(T)$, and even obscure the actual value.

Secondly, if $T > 1$ and $\Delta < 1$, the fact that (T/Δ) steps are required may create an exorbitant execution time. This can be serious consideration in connection with various "on-line" decision processes of the type that occur in weather prediction and medical diagnosis.

Thirdly, even if the stability problem is resolved and the time requirements are acceptable, the procedure nevertheless still possesses some esthetic handicaps. Often, the mathematical model of the original physical process is known to be rather "rough and ready". What is desired from the equation then, is a reasonable estimate of the functional values of a few grid points rather than any highly accurate determination of entire set of values $(w(k\Delta))$.

Here we wish to examine the general application of differential quadrature.

6.9 Differential Quadrature

Let the points $0 = t_0 < t_1 < t_2 < \ldots < t_N$ be selected and the coefficient matrix $A = (a_{ij})$ be chosen so that

$$(6.9.1) \qquad y'(t_i) \cong \sum_{j=1}^{N} a_{ij}\, y(t_j) \quad .$$

There are several ways of doing this based upon the method of least squares, methods akin to Gaussian quadrature, and the emerging theory of splines.

The equation of (6.9.1) then becomes

$$(6.9.2) \qquad \sum_{j=1}^{N} a_{ij}\, y(t_j) \cong g(y(t_i)) \ , \qquad i = 1,2,\ldots,N \quad .$$

We can now proceed in several ways. To begin with, we can consider the system of equations

$$(6.9.3) \qquad \sum_{j=1}^{N} a_{ij}\, y(t_j) = g(y(t_i)) \ , \qquad i = 1,2,\ldots,N \ ,$$

as a method of determining $y(t_i)$. Secondly, we can use a least squares technique. Thirdly, we can use a Chebyshev norm and apply linear and nonlinear programming techniques.

6.10 g(y) Linear

If $g(y) = By$, an application of the least squares technique leads to the problem of the solution of a linear system of equations. In a number of cases we can employ intrinsic properties of the physical process to determine a regularization, or "penalty", function to ensure a well-conditioned system.

If a Chebyshev norm is employed, linear programming techniques can be used.

6.11 g(y) Nonlinear

If $g(y)$ is nonlinear, the minimization problem associated with a least squares procedure requires some use of successive approximations. One way to obtain a good initial approximation is to use a low-order differential quadrature, where the minimization process is easy to carry out, plus interpolation.

6.12 Partial Differential Equations

The method can be applied to various classes of partial differential equations reducing them to ordinary differential equations and then to finite-dimensional systems. Consider, for example, the equation

$$u_t = g(u, u_x), \qquad u(x,0) = h(x) \quad . \tag{6.12.1}$$

Write

$$u_x \Big|_{x=x_i} = \sum_{j=1}^{N} a_{ij}\, u(x_j, t), \qquad i = 1,2,\ldots,N, \tag{6.12.2}$$

where $x_1 < x_2 < \ldots < x_N$, and consider the associated system of ordinary differential equations

$$(6.12.3) \qquad v_i' = g\left(v_i(t), \sum_{j=1}^{N} a_{ij}\, v_j(t)\right), \quad v_i(0) = h(x_i),$$

where $v_i \cong u(x_i, t)$, $i = 1,2,\ldots,N$.

We can eliminate the t derivative if desired with a repeated application of this procedure. Alternatively, a Bubnov-Galerkin technique can be used to find an approximation solution of (6.12.3).

6.13 Use of the Laplace Transform

Consider the equation

$$(6.13.1) \qquad x' = A(t)x, \quad x(0) = c.$$

If we take the Laplace transform of this equation, we have

$$(6.13.2) \qquad sL(x) - c = L(A(t)x).$$

The integrals are evaluated by quadrature. Thus, the original problem of solving a linear differential equation has been approximated by solving a system of linear equations. We shall give an application of this technique in Chapter 6 and present some numerical results.

6.14 Finding a Particular Solution

Let us now discuss the problem of obtaining the smaller solution. This is an important problem computationally since any amount of round-off error brings in the larger solution. This is a significant problem. We thus have computational instability.

We shall begin with the second-order linear differential equation

$$(6.14.1) \qquad u'' - a(t)u = 0.$$

Suppose we want to calculate the smaller solution. We can use the identity of Abel

$$\begin{vmatrix} u & u' \\ v & v' \end{vmatrix} = k \quad . \tag{6.14.2}$$

This means that if we know one solution, we can find the general solution by solving a first-order differential equation.

For the case of the second-order equation, we can also obtain the smaller solution by integrating backwards. The smaller solution is now the larger solution and computationally stable.

For higher order linear differential equations, we can use the device of integrating backwards to find the smallest solution. The problem of finding the largest and smallest solutions can be approached in various ways. In general, it will be necessary to combine some analysis with the computer. This is true in most cases. It is very seldom that we can use a computer without any mathematical analysis.

Consequently, having obtained the largest solution, we want a systematic way of lowering the order of the differential equation. Let us suppose that u is the largest solution. Consider the equation for uv. We see that this is an equation of one lower order for v'. We may now continue in this fashion to obtain all the solutions.

Bibliography and Comments

Section 6.3. This method was presented in, Bellman, R., "A Non-Singular Matrix is an Exponential", Journal of Mathematical and Physical Sciences, 1979.

Section 6.5. The result of Hermite is given in, Hermite, C., "Sur Quelques Applications des Fonctions Elliptiques", Oeuvres, Vol. 3, 1912, pp. 266-418.

Section 6.7. We are following the paper, Bellman, R. and Casti, J., "Differential Quadrature and Long-Term Integration", Journal of Math. Analysis and Appl., Vol. 34, No. 2, 1971, pp. 235-238.

Section 6.12. For some applications of this method, see Bellman, R., Kashef, B. G., and Casti, J., "Differential Quadrature: A Technique for the Rapid Solution of Nonlinear Partial Differential Equations", J. of Comp. Physics, Vol. 10, No. 1, Aug. 1972, pp. 40-52.

Chapter VII

NONLINEAR DIFFERENTIAL EQUATIONS

7.1 Introduction

Let us now turn to nonlinear differential equations. Consider the vector differential equation

$$(7.1.1) \qquad x' = g(x) , \qquad x(0) = c \quad .$$

Here each component of $g(x)$ is a power series in the components of x, lacking constant terms. Let us show that x depends upon both t and the initial value by writing $x(c,t)$. We have the condition $x(c,0) = c$.

We are interested in calculating a particular value, without calculating intermediate values. Theoretically, we know this is possible since this value is determined by the initial conditions. Computationally, this is of importance since there is always the danger of round-off error. We have this danger even in the linear case.

We shall approach this problem by four methods. One employs iteration, the second employs relative invariants, the third employs the Laplace Transform and the fourth employs Carleman linearization. This is consistent with what we have said before. We want to compute every important number by at least two methods. In most problems, as computers become faster, time is not a factor. However, accuracy always is. In a large calculation, we want to calculate important quantities by different methods and compare the results before proceeding.

7.2 Fundamental Semigroup Relation

We have

$$x(c,t + s) = x(x(c,s), t) \quad . \tag{7.2.1}$$

If we take t and s as integers, we see that we have a problem in iteration. It is well known that there is a close connection between iteration and analytic differential equations.

We can iterate in many different ways depending upon the units we use. It is not even necessary to use any integration since we can write the solution as a power series in t or c.

7.3 Acceleration of Iteration and Powers of Two

We can speed up this process considerably by using powers of two as usual. We can iterate once, and then iterate this function, and so on.

We see that some calculation is required. It is necessary to calculate the fundamental power series. This is consistent with what we have said before. For effective computation, we need a blend of mathematical analysis and the computer.

7.4 Relative Invariants

Let us now sketch a method that yields a great deal more information in dealing with analytic differential equations. As will be seen, this method is closely connected with iteration. Consider the scalar equation

$$\frac{du}{dt} = a_1 u + a_2 u^2 + \ldots, \qquad u(0) = c \ , \tag{7.4.1}$$

where $a_1 \neq 0$.

Let us see if we can determine a function of the solution u, $f(u)$, possessing the property that

(7.4.2) $$\frac{df}{dt} = \lambda f \quad .$$

Taking f to have the form

(7.4.3) $$f(u) = u + b_2 u^2 + \dots,$$

we obtain, upon combining (7.4.1) and (7.4.2),

$$\frac{df}{dt} = (1 + 2b_2 u + \dots) \frac{du}{dt}$$

(7.4.4) $$= (1 + 2b_2 u + \dots) (a_1 u + a_2 u^2 + \dots)$$

$$= \lambda(u + b_2 u^2 + \dots) \quad .$$

Equating coefficients of u, we see that $\lambda = a_1$, and that we can obtain a simple recurrence relation for the coefficients b_n.

Having thus determined $f(u)$, we return to (7.4.2) and note that we can solve explicitly,

(7.4.5) $$f(u) = e^{\lambda t} f(c) \qquad (\text{since } u(0) = c) \quad .$$

Thus u itself is given by

(7.4.6) $$u = f^{-1} (e^{\lambda t} f(c)) \quad .$$

If $Re(\lambda) < 0$, this yields the asymptotic behavior of u as $t \to \infty$. The same technique can be applied to the study of vector differential equations with the usual complications of multidimensionality.

Consider the two-dimensional system $dx/dt = Ax + g(x)$, $x(0) = c$, where A has distinct characteristic roots and $g(x)$ is analytic in components of x, lacking zero order and first-order terms. We can make a change of variable $x = Ty$ that has the effect of converting it into a system of the form

$$\frac{dy_1}{dt} = \lambda_1 y_1 + h_1(y_1, y_2) , \qquad y_1(0) = c_1 , \tag{7.4.7}$$

$$\frac{dy_2}{dt} = \lambda_2 y_2 + h_2(y_1, y_2) , \qquad y_2(0) = c_2 ,$$

where h_1 and h_2 are of second order or higher.

Let us assume that the characteristic roots are incommensurable. It is then easy to show we may solve for the coefficients of the relative invariants recurrently. Thus, the problem of finding the solution at a particular point has been converted into solving a system of equations.

7.5 An Alternate Approach

Consider the differential equation

$$x' = Ax + g(x) , \qquad x(0) = c \quad . \tag{7.5.1}$$

We assume that $g(x)$ is analytic in the components of x lacking constant and first degree terms. As we know, the differential equation can be converted into the integral equation

$$x = e^{At} c + \int_0^t e^{A(t - s)} g(x(s)) \, ds \quad . \tag{7.5.2}$$

We assume that the characteristic roots $\lambda_1, \lambda_2, \ldots, \lambda_n$, are distinct and that A has been reduced to diagonal form by a canonical transformation. The differential equation exists in an interval which depends upon c. If A is a stability matrix and if the norm of c is sufficiently small, the Poincaré-Lyapunov theorem tells us that the solution exists for all t.

Let us now solve the integral equation by iteration. If the incommensurability condition holds, we see that no powers of t will occur. We see the reason for the incommensurability condition if we proceed in this fashion.

The solution will then be power series in $e^{-\lambda_i t}$. If we now invert the series and solve for the $e^{-\lambda_i t}$ we obtain an equation

$$e^{-\lambda_i t} = f_i(x_1, x_2, \ldots, x_n) \quad . \tag{7.5.3}$$

The solution for a particular value of t is once again reduced to solving a set of simultaneous equations.

7.6 Relative Invariants in the Commensurable Case

Consider the two-dimensional system

$$\begin{aligned} u' &= -\lambda_1 u + \sum_{m+n\geq 2} a_{mn} u^m v^n , \qquad u(0) = c_1 , \\ v' &= -\lambda_2 v + \sum_{m+n\geq 2} b_{mn} u^m v^n , \qquad v(0) = c_2 , \end{aligned} \tag{7.6.1}$$

where $\lambda_1, \lambda_2 > 0$ and the two power series converge for $|u|, |v| \leq c_3$. If $|c_1|, |c_2|$ are sufficiently small, Poincaré-Lyapunov theory tells us that (7.6.1) has a unique solution which exists for $t \geq 0$, and which has the asymptotic behavior

$$u \sim g(c_1, c_2)\, e^{-\lambda_1 t} ,$$

$$v \sim h(c_1, c_2)\, e^{-\lambda_2 t} ,$$

where we assume that $\lambda_2 > \lambda_1$. The determination of the functions $g\,(c_1, c_2)$, $h\,(c_1, c_2)$ is perhaps most easily effected by means of invariant imbedding.

Once this result has been established, a number of further questions arise:

(a) How does one obtain more precise asymptotic results?

(b) How does one calculate $u(t)$ and $v(t)$ for a particular value t_1 without calculating values for $0 < t < t_1$?

(c) How does one calculate expected values monomials $u^m v^n$ in the case where c_1 and c_2 are random variables with a sufficiently small range?

If λ_1 and λ_2 are incommensurable, which is to say if $k\lambda_1 \neq l\lambda_2$ for any integers k and l, all three questions can be answered rather simply by means of relative invariants.

Specifically, we seek functions of the form

$$\phi(u, v) = u + \sum_{m+n\geq 2} c_{mn} u^m v^n ,$$

(7.6.3)

$$\Psi(u, v) = v + \sum_{m+n\geq 2} d_{mn} u^m v^n ,$$

where the power series are convergent for $|u|$, $|v|$ sufficiently small, with the property that

$$\frac{d\phi}{dt} = -\lambda_1 \phi ,$$

(7.6.4)

$$\frac{d\Psi}{dt} = -\lambda_2 \Psi ,$$

whenever u and v satisfy (7.6.1). We see the origin of the term "relative invariants".

If (7.6.4) holds, we have

$$\phi(u, v) = e^{-\lambda_1 t} \phi(c_1, c_2) ,$$

(7.6.5)

$$\Psi(u, v) = e^{-\lambda_2 t} \Psi(c_1, c_2) .$$

These representations furnish a direct way of answering the three questions posed in (7.6.3).

If we attempt to determine the coefficients in (7.6.3) in a formal fashion, we have upon substituting, the expressions in (7.6.3) into (7.6.4) and equating coefficients

(7.6.6) $$- \lambda_1 mc_{mn} - \lambda_2 nc_{mn} = - \lambda_1 c_{mn} + \dots$$

where the three dots represent terms in the coefficients c_{mn} and d_{mn} with smaller indices. Hence we can solve recurrently for the coefficients provided that at no stage

(7.6.7) $$\lambda_1 (m - 1) + \lambda_2 n \neq 0 \quad .$$

The question of convergence of the power series can be treated in a number of ways.

Consider the system

(7.6.8) $$\begin{aligned} u' &= - u , & u (0) &= c_1 , \\ v' &= - 2v + u^2 , & v (0) &= c_2 , \end{aligned}$$

where the condition of (7.6.7) is violated. Relative invariants of the foregoing type do not exist here. The solution of (7.6.8) is

(7.6.9) $$\begin{aligned} u &= c_1 e^{-t} , \\ v &= c_2 e^{-2t} + c_1^2 te^{-2t} \quad . \end{aligned}$$

Observe, however, that we can assert the existence of generalized relative invariants

(7.6.10) $$\begin{aligned} \phi (u, v) &= u , \\ \Psi (u, v) &= v - c_1^2 tu^2 \\ &= v + c_1^2 u^2 (\log u - \log c_1) \quad . \end{aligned}$$

Invariants of this non-analytic type are perfectly acceptable as far as providing answers to the questions of (7.6.3) are concerned. Let us then turn to the problem of obtaining these generalized relative invariants. We shall consider a particular case to illustrate the method.

Without loss of generality, we can take the basic system to be

(7.6.11)
$$\begin{aligned} u' &= -u, & u(0) &= c_1, \\ v' &= -2v + \sum_{m+n\geq 2} b_{mn} u^m v^n, & v(0) &= c_2. \end{aligned}$$

Then one relative invariant is clearly $\phi(u, v) = u$. Write

(7.6.12)
$$\Psi(u, v) = v + a_1 u^2 + a_2 u^2 \log u + \sum{}' a_{mn} u^m v^n ,$$

where the prime indicates that $m + n \geq 2$, and the value $m = 2$, $n = 0$ is omitted. Then

(7.6.13)
$$\begin{aligned} \frac{d\Psi}{dt} &= v' + 2a_1 uu' + 2a_2 (u \log u)\, u' + a_2 uu' + \ldots \\ &= (-2v + b_{20} u^2 + \ldots) - 2a_1 u^2 - 2a_2 u^2 \log u - a_2 u^2 + \ldots \end{aligned}$$

Hence, if we postulate the relation

(7.6.14)
$$\frac{d\Psi}{dt} = -2\Psi$$

we must have upon equating the coefficients of u^2,

(7.6.15)
$$-2a_1 - a_2 + b_{20} = -2a_1 ,$$

or

(7.6.16)
$$a_2 = b_{20} .$$

Observe that this leaves a_1 undetermined. This should be the case since

(7.6.17)
$$\frac{d}{dt}(a_1 u^2) = -2(a_1 u^2)$$

for any value of a_1 which means that we can modify any particular $\Psi(u, v)$ by adding a term $a_1 u^2$. The remaining coefficients in $\Psi(u, v)$ are obtained in the usual fashion.

We shall not discuss the convergence of the series for Ψ here. Let us, however, examine the use of the relative invariants

$$\phi(u, v) = u \quad ,$$
(7.6.18)
$$\Psi(u, v) = v + a_1 u^2 + a_2 u^2 \log u + \Sigma' a_{mn} u^m v^n$$

to answer the question in (7.6.3). The nonanalytic form of $\Psi(u, v)$ causes some difficulties as far as solving for u and v is concerned in the relations

$$\phi(u, v) = \phi(c_1, c_2) e^{-t} \quad ,$$
(7.6.19)
$$\Psi(u, v) = \Psi(c_1, c_2) e^{-2t} \quad .$$

To avoid this difficulty, we can replace $\log u$ by $-t + \log c_1$ and then use any of a number of methods for the reversion of series. This can, for example, be carried out efficiently by means of the Lagrange expansion.

7.7 Nonlinear Differential Equations

Let us now turn our attention to some nonlinear differential equations of conventional type. Our aim is to show that a combination of the Laplace Transform and various methods of successive approximations yields feasible techniques for obtaining numerical solutions.

Let us begin with the simple differential equation

(7.7.1) $$u' = -u - u^2 , \qquad u(0) = c \quad .$$

where c is taken to be non-negative and, at this stage of the game, small, for example; c is of the order of magnitude of 0.1 or 0.5. That the equation possesses a simple explicit solution is of no matter to us. Taking Laplace Transforms, we have

$$\text{(7.7.2)} \qquad L(u') = -L(u) - L(u^2) \quad ,$$

or, in the usual fashion,

$$\text{(7.7.3)} \qquad L(u) = \frac{c - L(u^2)}{(s+1)} \quad .$$

We will solve this nonlinear integral equation numerically by way of successive approximations combined with numerical quadrature. Let $u_0(t)$ be the solution of

$$\text{(7.7.4)} \qquad L(u_0) = \frac{c}{s+1} \quad .$$

In this case, we see that $u_0 = ce^{-t}$, but this explicit analytic representation is of no importance to us here since we are interested in general procedures. We employ a quadrature technique to obtain the values $\{u_0(t_i)\}$, $i = 1, 2, \ldots, N$.

Next, we determine u_1 by means of the equation

$$\text{(7.7.5)} \qquad L(u_1) = \frac{c - L(u_0^2)}{s+1} \quad .$$

To evaluate the expression $L(u_0^2)$, we employ the same quadrature formula

$$\text{(7.7.6)} \qquad L(u_0^2) = \sum_{i=1}^{N} w_i u_0^2 (t_i) r_i^{s-1} \quad .$$

Since the quantities $u_0^2(t_i)$ have already been determined, we can employ an inversion method to determine $\{u_1(t_i)\}$, $i = 1, 2, \ldots, N$.

Continuing in this way, we determine the values $\{u_n(t_i)\}$ by means of the relation

$$\text{(7.7.7)} \qquad L(u_n) = \frac{c - L(u_{n-1}^2)}{(s+1)} \quad .$$

Naturally, we expect that this procedure will converge only for $|c|$ sufficiently small.

In Figs. 7.1 and 7.2, we compare the results of direct numerical integration of the differential equation with the integration technique described above.

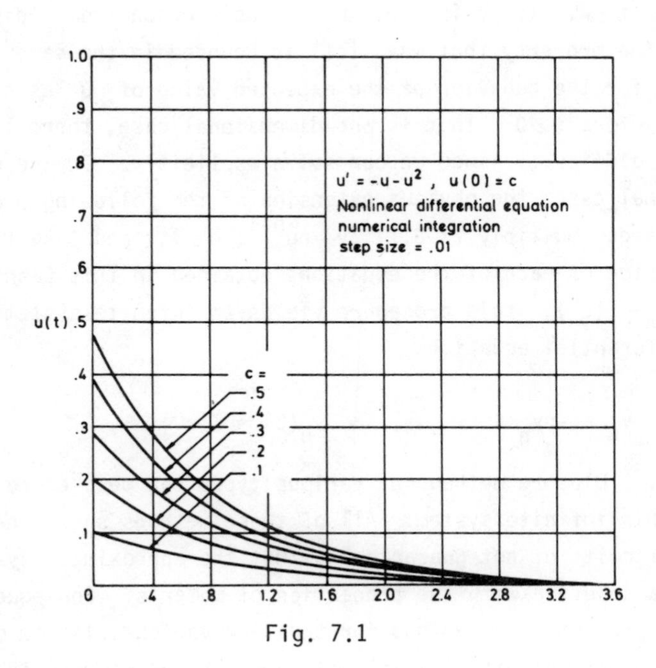

Fig. 7.1

Fig. 7.2

7.8 Stochastic Differential Equations

If $|c|$ is sufficiently small, the solution of the nonlinear differential equation

$$(7.8.1) \qquad u' = -u + u^2 , \qquad u(0) = c$$

exists for $t \geq 0$. If c is a random variable drawn from a distribution with the property that max $|c|$ is bounded in the same fashion, we can ask for the behavior of the expected value of u as a function of t for $t \geq 0$. In this one-dimensional case, there is, of course, no difficulty since we can solve explicitly. In the general N-dimensional case, the obvious extension of the following procedure is often used. Multiply (7.8.1) by nu^{n-1}, $n \geq 1$, and take the expected value of each of the equations obtained in this fashion. Writing $v_n = E(u^n)$, this procedure yields an infinite system of linear differential equations

$$(7.8.2) \qquad v_n' = -nv_n + nv_{n+1} , \qquad v_n(0) = E(c^n) ,$$

$n = 1, 2, \ldots$. Closure methods of various types can be used to truncate this infinite system. All of these methods share the common difficulty of not guaranteeing that the approximate system possesses a solution with the properties of moments. Consequently, we wish to present a new method for treating various classes of functional equations with initial values which are stochastic variables, and to illustrate this method in connection with the Burgers' equation,

$$(7.8.3) \qquad u_t + uu_x = \epsilon u_{xx} ,$$

$t > 0$, $0 \leq x \leq 2\pi$, $u(x, 0) = g(x)$, with $u(x, t)$ periodic of period 2π, and $g(x)$ a random periodic function. The way in which this randomness is attained, is described below. Since this equation can also be integrated in reasonably explicit analytic terms, it provides an excellent test for analytic approximations and computational techniques and has frequently been used for this purpose.

7.9 Relative Invariants

Let us now apply the technique of relative invariants. Let $\phi(u)$ be an analytic function of u in some neighborhood of $u = 0$, $\phi(u) = u + \sum_{k=2}^{\infty} b_k u^k$, and let the coefficients be determined by the condition that

$$\frac{d}{dt}\phi(u) = -\phi(u) \quad . \tag{7.9.1}$$

Writing

$$\begin{aligned} \frac{d}{dt}\phi(u) &= (u + b_2u^2 + b_3u^3 + \ldots) \\ u' &= (u + b_2u^2 + \ldots)(-u + u^2) \quad , \end{aligned} \tag{7.9.2}$$

and equating coefficients of powers of u, we see that the coefficients can be determined recurrently (i.e., b_k can be obtained in terms of $b_2, b_3, \ldots, b_{k-1}$). The fact that the series obtained in this way has a nonzero radius of convergence can readily be established using the classical results of the theory of iteration. Regarding Eq. (7.9.1) as a differential equation for $\phi(u)$, we see that $\phi(u) = e^{-t}\,\phi(c)$. From this expression for $\phi(u)$, it is immediate that the generalized moments $\exp(\phi(u)^k)$ can be readily obtained. Writing $u = \phi^{-1}(e^{-t}\phi(c))$, and expanding, using the Lagrange expansion or otherwise, we have

$$u = e^{-t}\phi(c) + a_2\, e^{-2t}\,\phi(c)^2 + \ldots \quad . \tag{7.9.3}$$

From this, we obtain the ordinary moments of u. It is important to note that the determination of the coefficient sequences $\{a_n\}$, $\{b_n\}$ is independent of the statistics of c. The case where there is no term in u is very much more delicate and we shall avoid any discussion here.

7.10 Burgers' Equation

Finite dimensional extensions of the following procedure exist, and we shall indicate it is easy to obtain infinite dimensional analogues. Write $u(x, t) = \sum_{n=-\infty}^{\infty} u_n(t) \; \phi_n(x)$ where $\phi_n(x) = \exp(inx/2\pi)$. Then we obtain the infinite-dimensional system of ordinary differential equations,

$$(7.10.1) \qquad u_n' + \varepsilon n^2 u_n = -i \sum_{k=-\infty}^{\infty} k \, u_k(t) \, \phi_{n-k}(t) \quad .$$

Let the initial conditions be $u_n(\emptyset) = g_n$, where the random variables g_n are the Fourier coefficients of the random function $g(x)$. By analogy with what has preceded, we seek a function of the form

$$(7.10.2) \qquad \phi_n(u) = \phi_n(u_0, u_1, u_{-1}, \ldots) \; .$$

$$= u_n + \sum_{m,k} a_{mk} \, u_m \, u_k + \sum_{k,m,l} a_{kml} \, u_k \, u_m \, u_l + \cdots$$

which satisfies the differential equation

$$(7.10.3) \qquad \frac{d}{dt} \phi = -\varepsilon n^2 \phi \quad .$$

The coefficients a_{mk}, a_{kml}, etc., depend on n. Equating coefficients in (7.10.3) as before, we see once again that they can be obtained recurrently. As before, the coefficients are independent of the statistics of the g_n. In particular

$$(7.10.4) \qquad a_{mk} = \frac{-ik}{\sqrt{2\pi}} \; \frac{\delta(n-m-k)}{\varepsilon(m^2 + k^2)} \quad ;$$

where $\delta(i)$ is the Kronecker delta symbol. We then have, upon integration,

$$\phi_n(u) = e^{-n^2 \varepsilon t} \; \phi_n(g_0, g_1, \ldots) \quad .$$

From this, we can find expansions for the moments of the u_n, and for other expected values and correlations.

7.11 Carleman Linearization

It will be sufficient to consider a simple example. It will be clear that the method can be used in more general cases.

Consider the differential equation

$$\frac{du}{dt} = -u + u^2 , \qquad u(0) = c \quad . \tag{7.11.1}$$

Consider the simple result

$$\frac{du^n}{dt} = n\, u^{n-1} \frac{du}{dt} \quad . \tag{7.11.2}$$

Using this result, the original nonlinear equation is converted into a linear equation of infinite order.

7.12 Truncation

In the foregoing section, we have shown how a non-linear equation may be converted into a linear equation of infinite order.

In order to treat this equation, it is necessary to truncate in some fashion.

Once we have done so, we can use the method given in Chapter 4.

There are many ways of performing this truncation.

Every method of truncation leads to questions of stability. It will be seen how many open questions there are.

7.13 Discussion

If we analyse the linearization technique given above we see that we have used the following two facts:

1. The derivative of any member of the sequence is a linear combination of other members of the sequence.
2. The product of two members of the sequence is a linear combination of members of the sequence.

These conditions are satisfied by orthogonal polynomials for example. Hence, if we have some idea of the range of the solution, we can do a lot better than above. For example, we may want to use Chebyshev polynomials.

7.14 Solutions which Exist for All t

Very precise results concerning asymptotic behavior can be obtained if we assume we are dealing with solutions which exist for all t. Consider the polynomial equation

$$(7.14.1) \qquad p(u, u', t) = 0 \quad .$$

The asymptotic behavior was obtained by Borel, Hardy, and Lindelof.

The equation

$$(7.14.2) \qquad u'' - t^m u^n = 0$$

was studied for all values of the parameters by Fowler. This equation is usually known as the Emden-Fowler-Fermi-Thomas equation. It arose in the researches of Emden in astrophysics and of Fermi-Thomas in quantum mechanics. Another interesting class of solutions which exists for all time are peridoic solutions of nonlinear second-order differential equations. Their behavior is quite complex. For example, the nonlinear spring

$$(7.14.3) \qquad u'' + u + ku^3 = 0 \quad ,$$

possesses periodic solutions which can be obtained from the linear equation. On the other hand, the Van der Pol equation, the equation of the multivibrator

$$(7.14.4) \qquad u'' + \lambda(u^2 - 1)u' + u = 0 \quad .$$

has a single periodic solution.

If we allow periodic forcing terms, we obtain even more complex behavior.

An analog of periodic solutions for higher dimensions is a periodic surface. Not much is known in this area.

Bibliography and Comments

Section 7.1. We are following the paper, Bellman, R., "Selective Computation - VIII. Nonlinear Differential Equations, Iteration, and Relative Invariants", Journal of Nonlinear Analysis: Theory, Methods and Appl., Vol. 4, No. 1, 1980, pp. 71-72.

For the theory of iteration, see Bellman, R., Methods of Nonlinear Analysis, Vol. II, Academic Press, Inc., New York, 1973.

Section 7.6. We are following the paper, Bellman, R., "Relative Invariants in the Commensurable Case", Journal of Mathematical Analysis and Applications, Vol. 28, No. 2, Nov. 1969, pp. 400-404.

Section 7.8. We are following the paper, Bellman, R., and Richardson, J. M., "Relative Invariants, Closure, and Stochastic Differential Equations", Journal of Mathematical Analysis and Applications, Vol. 12, May 1966, pp. 294-296.

Section 7.11. We are following the paper, Bellman, R., "Selective Computation - IV. Nonlinear Equations and Carleman Linearization", Journal of Nonlinear Analysis: Theory, Methods and Appl., Vol. 3, No. 4, 1979, pp. 515-516.

Section 7.14. For polynomial differential equations of the first-order it was shown by Poincaré that the solution was given by automorphic functions. With digital computers now available, this seems to be an effective numerical method.

The results of Borel-Hardy-Lindel and Fowler are given in Bellman, R., Stability Theory of Differential Equations, Dover Reprint, New York, 1969.

For a rigorous derivation of the perturbation series, and for a topological discussion of periodic solutions in general, see Cesari, L., Asymptotic Behavior and Stability Problems in Ordinary Differential Equations, Springer-Verlag, Berlin, 1959.

Lefschetz, S., Lectures on Differential Equations, Princeton University Press, Princeton, New Jersey, 1946.

Stoker, J. J., Nonlinear Vibrators in Mechanical and Electrical Systems, Interscience Publishers, Inc., New York, 1950.

Poincaré, H., Methodes Nouvelles de la Mecanique Celeste, Vols. I, II, III, Gauthier-Villars, Paris, 1892.

Mori, R., Analytical Design of Vacuum Tube Blocking Oscillator, Electrotechnical Laboratory, No. 616, Tokyo, 1961.

The preceding results have a great deal of significance as far as quantum mechanics is concerned, as Minorsky and the author discussed in conversations in 1946. They show that an infinitesimal amount of nonlinearity both quantifies phase space and determines the amplitude of periodic motion, which is to say of orbits. Hence, the stable limit cycles (as the periodic solutions are called) can represent stable states. The presence of interlarded unstable limit cycles means that energy is required to perturb the orbit from one stable state to another. The extreme stability of the stable cycles explains why no transit time is observed, and perhaps dissipates a certain amount of mysticism about instantaneous change of state.

These ideas have recently gained prominence. See Andrade e Silva, Joao, Fer Francis, Lenuste Philippe, Lochak, Georges, "Non-linéarite, Cycles Limites et Quantification", C. R. Acad. Sci. Paris, 251, 1960, pp. 2662-2664,

and

Duerr-Heisenberg-Mitter-Schlieder-Yamazaki, "Zur Theorie der Elementarteilchen", Zeit Fur Naturforsch, Bd. 14A, 1959, pp. 441-485.

Chapter VIII
ASYMPTOTIC BEHAVIOR

8.1 Introduction

The use of principles of invariance, as in invariant imbedding and dynamic programming, leads characteristically to functional equations of the form

$$(8.1.1) \qquad f_{n+1}(p) = T_n(f_n(g(p))) , \qquad n = 0, 1, 2, \ldots ,$$

where $f_0(p)$ is known. The computational solution proceeds stagewise, with f_1 determined from a knowledge of f_0, f_2 determined by f_1, and so on.

In general, what is desired is the transient behavior, small n, and the steady-state, or asymptotic behavior as $n \to \infty$. In a number of significant processes — radiative transfer, control theory, inventory theory and Markovian decision processes in general — only the asymptotic results are of interest. This is also the case in the application of gradient techniques.

In some cases where the asymptotic results are not of primary importance, they are worth obtaining in order to test the accuracy of the numerical techniques, since the steady-state solution can often be derived by other independent means.

It is evident that direct step-by-step calculation of the asymptotic solution, using (8.1.1), is time consuming. Hence, it is important to develop extrapolation techniques. A most important step in this direction is the work of Shanks, closely related to the QD-algorithm of Rutishauser. We shall discuss below some extensions related to nonequally spaced observations or calculations. In this connection, let us mention the work of Kantorovich and Krylov concerning the improvement of convergence of Fourier series and other types of sequences.

In this chapter, we shall outline the application of nonlinear summability techniques to radiative transfer.

8.2 Motivation of Method

The fundamental assumption is that we possess an asymptotic expansion of the form

$$f_n \quad f_\infty + \sum_k a_k e^{-\lambda_k n} \tag{8.2.1}$$

as $n \to \infty$, or that we can sensibly approximate to f_n by an expression of the type appearing on the right-hand side. From theoretical considerations in invariant imbedding, dynamic programming, gradient techniques, and elsewhere, we obtain rigorous demonstrations of this relation. What is remarkable, as shown by Shanks, is that quite accurate results are obtained even when an asymptotic relation of this type does not hold, and even more remarkable is the fact that calculations based on small values of n give excellent estimates for f_∞.

If we take f_n to have the simple form

$$f_n = f_\infty + ae^{bn} \quad , \tag{8.2.2}$$

an immediate calculation yields

$$f_\infty = \frac{\begin{vmatrix} f_n & f_{n+1} \\ f_{n+1} & f_{n+2} \end{vmatrix}}{(f_n + f_{n+2} - 2 f_{n+1})} \quad . \tag{8.2.3}$$

In the next section we shall describe some experiments with this simple nonlinear predictor. As mentioned above, many further results based upon more sophisticated approximations will be found in Shanks.

8.3 Radiative Transfer

Let parallel rays of radiation be incident upon a plane-parallel slab of finite thickness which absorbs radiation and scatters it isotropically (see Fig. 7.3).

Using the theory of invariant imbedding, we obtain an equation for the diffuse reflection $r(\theta, \Psi, x)$ which leads to a feasible computational algorithm. As pointed out above, it is of interest to obtain the results for infinite x in this way in order to compare with previous results of Ambarzumian and Chandrasekhar obtained in another fashion.

If the albedo for single scattering is 0.9, a thickness of 6 mean free paths is required to saturate, i.e., to obtain a reflection coefficient equivalent to infinite thickness to about four decimal places. For the particular case of input and output angles to 60 degrees to the normal, we calculated the reflection coefficient at thickness of 0.00, 0.02, 0.04,..., 1.20 mean free paths. These values are listed in Table 8.1, which is read across from left to right in each row and from the top row to the bottom row. For a thickness of 1.2 mean free paths the calculated value is 0.23295887.

Next we use (8.2.3) on each set of three consecutive entries in Table 8.1 to produce the entries in Table 8.2. These are predictions of the limiting value, which is 0.272389.

Using the predictions of Table 8.2, we can use the formula once again to produce another set of predicted values. These are shown in Table 8.3.

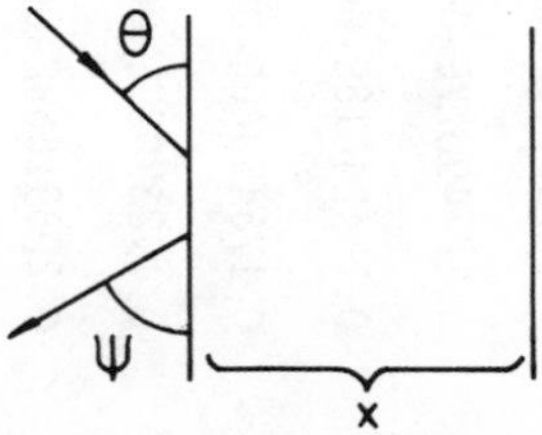

Fig. 7.3

Table 8.1

THE SET OF REFLECTION COEFFICIENTS

0.	0.90203642E-02	0.17958777E-01	0.26708581E-01	0.35219225E-01	0.43468159E-01
0.51446833E-01	0.59153822E-01	0.66591462E-01	0.73764186E-01	0.80677649E-01	0.87338272E-01
0.93752968E-01	0.99928968E-01	0.10587369E-00	0.11159468E-00	0.11709950E-00	0.12239571E-00
0.12749077E-00	0.13239207E-00	0.13710687E-00	0.14164225E-00	0.14600514E-00	0.15020226E-00
0.15424015E-00	0.15812514E-00	0.16186335E-00	0.16546070E-00	0.16892288E-00	0.17225537E-00
0.17546348E-00	0.17855225E-00	0.18152658E-00	0.18439115E-00	0.18715045E-00	0.18980878E-00
0.19237027E-00	0.19483887E-00	0.19721837E-00	0.19951241E-00	0.20172445E-00	0.20385781E-00
0.20591568E-00	0.20790109E-00	0.20981697E-00	0.21166609E-00	0.21345112E-00	0.21517462E-00
0.21683901E-00	0.21844663E-00	0.21999974E-00	0.22150043E-00	0.22295079E-00	0.22435274E-00
0.22570816E-00	0.22701886E-00	0.22828654E-00	0.22951284E-00	0.23069934E-00	0.23184753E-00
0.23295887E-00					

Table 8.2

THE FIRST ITERATION

0.99287306E-00	0.43262226E-00	0.33807480E-00	0.30347078E-00	0.28699413E-00	0.27778906E-00
0.27197126E-00	0.26796789E-00	0.26503253E-00	0.26280121E-00	0.26107164E-00	0.25972616E-00
0.25868013E-00	0.25788119E-00	0.25728158E-00	0.25684950E-00	0.25655429E-00	0.25637558E-00
0.25629469E-00	0.25628930E-00	0.25635426E-00	0.25647131E-00	0.25663887E-00	0.25683594E-00
0.25707100E-00	0.25732730E-00	0.25760114E-00	0.25789392E-00	0.25819670E-00	0.25850921E-00
0.25882599E-00	0.25915074E-00	0.25947401E-00	0.25979807E-00	0.26012429E-00	0.26044731E-00
0.26076716E-00	0.26108547E-00	0.26139567E-00	0.26170678E-00	0.26200978E-00	0.26230683E-00
0.26260211E-00	0.26288666E-00	0.26316931E-00	0.26344289E-00	0.26371384E-00	0.26397514E-00
0.26423369E-00	0.26448102E-00	0.26472771E-00	0.26496568E-00	0.26519617E-00	0.26542251E-00
0.26564503E-00	0.26585729E-00	0.26606464E-00	0.26626956E-00	0.26646777E-00	

Table 8.3

FURTHER SET OF PREDICTED VALUES

0.31887970E-00	0.28349463E-00	0.27201789E-00	0.26609017E-00	0.26206234E-00	0.25909086E-00
0.25696488E-00	0.25572946E-00	0.25510980E-00	0.25501280E-00	0.25502606E-00	0.25529800E-00
0.25547789E-00	0.25573519E-00	0.25591746E-00	0.25610149E-00	0.25622779E-00	0.25628892E-00
0.25629428E-00	0.25620828E-00	0.25608309E-00	0.25551960E-00	0.25561663E-00	0.25423604E-00
0.25332305E-00	0.25336742E-00	0.24902849E-00	0.24848599E-00	0.23524898E-00	0.24591999E-00
0.33054144E-00	0.12678152E-00	0.21058628E-00	0.29312269E-00	0.29283288E-00	0.32871063E-00
0.27323254E-00	0.15552984E-00	0.27332580E-00	0.27714357E-00	0.31166675E-00	0.27043446E-00
0.30527262E-00	0.27169371E-00	0.29164530E-00	0.27104514E-00	0.28857920E-00	0.26993456E-00
0.36083987E-00	0.27145021E-00	0.27231190E-00	0.27770513E-00	0.27866112E-00	0.27024904E-00
0.27479445E-00	0.28370491E-00	0.27231149E-00			

We see that we may predict a limiting value of about 0.27, which is quite accurate enough for many purposes.

Also remarkable is the fact that a double precision calculation of values at 0.005, 0.010,..., 0.300, with $r(\pi/3, \pi/3, 0.300) = 0.111595$ (which is less than 50% of the limiting value), predicts a limiting value of 0.27.

It is clear what a great saving in time can be obtained in this way.

The questions of which predictor formula to use, what increment of thickness to employ, and how many increments to use in order to predict the limiting values most efficiently, are still open.

8.4 Time-Dependent Processes

Time-dependent radiative-transfer problems may be resolved computationally in a multistage fashion. First we use invariant imbedding to obtain a set of nonlinear partial differential-integral equations. A Laplace Transform then reduces these to equations identical in form to those encountered in the time-independent problem. These are integrated numerically for appropriately chosen values of the transform variable. Finally, we use a numerical inversion of the Laplace Transform. A disadvantage of this method is that the values of the desired function, say $u(t)$, are obtained at irregularly spaced points t_i, $i = 1, 2, \ldots, M$, and it is not easy to make t_m large. Hence, an accurate extrapolation method would be quite useful in this case.

Instead of requesting a representation of the type appearing in (8.2.1), we can ask that $u(t)$ be approximated to by a function $w(t)$ satisfying the linear differential equation

$$w^{(n)} + b_1 w^{(n-1)} + \ldots + b_n w = b_{n+1} \quad . \tag{8.4.1}$$

The unknown constants b_i and the initial values $w^{(i)}(0) = c_i$, $i = 0, 1, \ldots, n-1$, are to be determined by the condition that

$$\sum_{i=1}^{M} (u(t_i) - w(t_i))^2 \tag{8.4.2}$$

is a minimum. The numerical solution of problems of this type can be carried out quite easily.

8.5 Gradient Techniques

The general idea of the gradient technique is to solve an equation of the form $T(u) = 0$ by imbedding it within the solutions of

$$\frac{\partial u}{\partial t} = T(u) \quad . \tag{8.5.1}$$

The solutions of the original equation are taken as the steady-states of (8.5.1). Since in this case only the values at $t = \infty$ are desired, nonlinear summability results will save a good deal of computing time. Combined with quasilinearization, very accurate results can be obtained.

8.6 Asymptotic Behavior of Solutions to Initial Value Problems

The method of invariant imbedding has been under intensive study for several years, primarily as a device for the formulation of certain problems in various areas of physics — transport theory, wave propagation, transmission line studies, etc. The general ideas involved go back at least to Stokes. While it has been realized that some basic mathematical structure probably lies behind all of these diverse applications, it is only recently that substantial progress has been made in understanding these fundamentals. Results suggest that the general imbedding technique may be applied in a valuable way toward the reformulation and understanding of many mathematical problems, quite independent of any physical applications or concepts.

It is our purpose here to apply the invariant imbedding method to the following type of problem. Suppose one knows the behavior for large t of the fundamental solutions to the equation

$$Lu(t) = f(t)\,u(t) \tag{8.6.1}$$

where, to be specific, we assume L is a second-order linear differential operator. We seek the asymptotic behavior of that solution satisfying initial conditions

$$u(0) = c_1 , \qquad u'(0) = c_2 \quad . \tag{8.6.2}$$

It is obvious that knowledge of the asymptotic behavior of the linearly independent solutions of (8.6.1) is of little direct aid in solving the problem posed. The question is one which arises often in wave mechanics, where

$$Lu = \frac{d^2u}{dt^2} + u \quad . \tag{8.6.3}$$

Solutions are known. Another case of interest is

$$Lu = \frac{d^2u}{dt^2} - u \quad , \tag{8.6.4}$$

which has also been studied.

We shall see that the imbedding method provides a uniform device for dealing with (8.6.3) and (8.6.4), both of which we shall study in detail. It will become apparent from the investigation of these two specific cases that the method can be applied to much more general linear operators L, such as

$$\frac{d^2u}{dt^2} + u = f(t)u \quad . \tag{8.6.5}$$

8.7 A Representation of the Solution for Large t

We shall suppose that $f(t)$ is continuous and bounded on $0 \leq t < \infty$ and that

$$(8.7.1) \qquad \int_0^\infty |f(t)|dt < \infty \quad .$$

(The conditions on f are somewhat more restrictive than necessary. Indeed, cases in quantum mechanics in which f becomes infinite at $t = 0$ are not covered. We choose to leave necessary modifications to the reader).

Let $u(t) = u(t; x, \theta)$ be the solution of

$$(8.7.2a) \qquad \frac{d^2u}{dt^2} + u = f(t)\, u \quad ,$$

$$(8.7.2b) \qquad u(x) = u(x; x, \theta) = \cos\theta \quad ,$$

$$(8.7.2c) \qquad \frac{du}{dt}\Big|_x = u'\,(x; x, \theta) = \sin\theta \quad .$$

(We shall write $du/dt = u'(t; x, \theta)$.)

It is well known that there are two fundamental solutions to (8.7.2a) of the form, for large t,

$$(8.7.3) \qquad \begin{aligned} u_1(t) &= \sin t + o(1) \quad , \\ u_2(t) &= \cos t + o(1) \end{aligned}$$

so that the system (8.7.2) has the solution

$$(8.7.4) \qquad u(t; x, \theta) = A(x, \theta) \cos(t - x - \theta - \Psi(x, \theta)) + o(1) \quad .$$

Our problem, then, is to find the quantities $A(x, \theta)$, the amplitude, and $\Psi(x, \theta)$, the phase shift. We must first establish some of their properties.

It is convenient to replace (8.7.2) with the equivalent

ıtegral equation

$$u(t;\ x,\ \theta) = \cos(t - x - \theta) + \int_x^t \sin(t - w)\ f(w)\ u(w;\ x,\ \theta)\ dw \quad . \tag{8.7.5}$$

he solution to (8.7.2) or, equivalently, to (8.7.5), is known to xist and to be unique for all $t \geq 0$. It will be desirable, however, o consider $u(t;\ x,\ \theta)$ as defined only for $t \geq x$. Thus x is ruly the initial value of t. It also follows from classical heory that $u_x = \partial u/\partial x$ and $u_\theta = \partial u/\partial \theta$ exist and are continuous.

An inequality of Gronwall will be used repeatedly. For onvenience we state it here.

Gronwall's Inequality

Let $g(z)$ and $h(z)$ be continuous and non-negative for $\geq z_1$. Suppose that for some $c > 0$

$$g(z) \leq c + \int_{z_1}^{z} g(z')\ h(z')\ dz' \quad . \tag{8.7.6}$$

hen

$$g(z) \leq c \exp\left\{\int_{z_1}^{z} h(z')\ dz'\right\} \quad . \tag{8.7.7}$$

e can now prove our first result.

LEMMA 1. The solution $u(t;\ x,\ \theta)$ of (8.7.2) or (8.7.5) atisfies

$$|u(t;\ x,\ \theta)| \leq \exp\left(\int_x^t |f(w)|dw\right) \leq M \quad , \tag{8.7.8}$$

here M is a constant dependent only upon f.

PROOF. From (8.7.5)

$$|u(t;\, x,\, \theta)| \leq 1 + \int_x^t |f(w)|\; |u(w;\, x,\, \theta)|dw \quad .$$

An application of the Gronwall's inequality, together with condition (8.7.1) furnishes our result.

In view of (8.7.8) we may now write still another equation for u:

$$\begin{aligned} u(t;x,\theta) &= \cos\,(t-x-\theta) + \int_x^\infty \sin\,(t-w)\; f(w)\; u(w;x,\theta)\; dw \\ &\quad - \int_t^\infty \sin(t-w)\; f(w)\; u(w;x,\theta)\; dw \\ &= \sin t \left\{ \sin\,(x+\theta) + \int_x^\infty \cos w\; f(w)\; u(w;x,\theta)\; dw \right. \\ &\quad + \cos t \left\{ \cos\,(x+\theta) - \int_x^\infty \sin w\; f(w)\; u(w;x,\theta)\; dw \right. \\ &\quad - \int_x^\infty \sin(t-w)\; f(w)\; u(w;x,\theta)\; dw \\ &= B(x,\theta)\sin t + C(x,\theta)\cos t + \zeta(t;x,\theta) \quad . \end{aligned} \tag{8.7.9}$$

LEMMA 2. $B(x,\theta)$ and $C(x,\theta)$ have continuous partial derivatives with respect to both x and θ. Furthermore,

$$\lim_{t\to\infty} \zeta(t;x,\theta) = 0$$

uniformly in x and θ.

PROOF. That $\lim_{t\to\infty} \zeta(t;x,\theta) = 0$ uniformly follows readily from (8.7.1) and Lemma 1.

To study the properties of B and C we first note from

(8.7.5) and the known differentiability of u that

$$\frac{\partial u}{\partial x} = \sin(t-x-\theta) - \sin(t-x)\, f(x)\, u(x;x,\theta) + \int_x^t \sin(t-w)\, f(w)\, u_x(w;x,\theta)\, dw \quad . \tag{8.7.10}$$

Since $u(x;x,\theta) = \cos\theta$ and since $f(x)$ is assumed bounded, another application of Gronwall's inequality shows that

$$\left|\frac{\partial u}{\partial x}(t;x,\theta)\right| \le M_1 \quad , \tag{8.7.11}$$

where M_1 is independent of t, x, and θ. Thus we may write, using the definition of $B(x,\theta)$ (see (8.7.9)),

$$\frac{\partial B}{\partial x} = \cos(x+\theta) - \cos x \cos\theta\, f(w) + \int_x^\infty \cos w\, f(w)\, u_x(w;x,\theta)\, dw \quad , \tag{8.7.12}$$

the differentiation under the integral sign being justified by the uniform convergence of the integral. The continuity of $\partial B/\partial x$ follows easily from (8.7.12). Differentiability with respect to θ is even easier and we shall not pursue the details; treatment of $C(x,\theta)$ is completely analogous. This completes the proof of Lemma 2.

THEOREM 1. The solution $u(t;x,\theta)$ may be written in the form

$$u(t;x,\theta) = A(x,\theta)\cos(t-x-\theta-\Psi(x,\theta)) + \zeta(t;x,\theta) \quad , \tag{8.7.13}$$

where $A(x,\theta)$ and $\Psi(x,\theta)$ have continuous first partial derivatives with respect to x and θ and where

$$\lim_{x\to\infty} A(x,\theta) = 1 \quad , \tag{8.7.14}$$

$$\lim_{x\to\infty} \Psi(x,\theta) = 0 \quad .$$

PROOF. We must first demonstrate that $B(x,\theta)$ and $C(x,\theta)$ cannot both be zero. From the result (8.7.3) we know that the solution $u(t;x,\theta)$ may be written, for some B and C,

$$(8.7.15)\qquad u(t;x,\theta) = Bu_1(t) + Cu_2(t)$$

$$= B \sin t + C \cos\cdot t + O(1) \quad .$$

Comparing this with (8.7.9)

$$(8.7.16)\qquad (B(x,\theta) - B) \sin t + (C(x,\theta) - C) \cos t = O(1) \quad .$$

Thus $B(x,\theta) = B$ and $C(x,\theta) = C$. But if $B(x,\theta) = C(x,\theta) = 0$ then, from (8.7.15), $u(t;x,\theta) = 0$. This is a contradiction, and we conclude $B(x,\theta)$ and $C(x,\theta)$ cannot simultaneously vanish.

Hence we can write in the usual way

$$(8.7.17)\qquad u(t;x,\theta) = \sqrt{B^2(x,\theta) + C^2(x,\theta)}$$

$$\left\{ \frac{B(x,\theta)}{\sqrt{B^2(x,\theta) + C^2(x,\theta)}} \sin t \right.$$

$$\left. + \frac{(C(x,\theta)}{\sqrt{B^2(x,\theta) + C^2(x,\theta)}} \cos t \right\} + \zeta(t;x,\theta)$$

$$= A(x,\theta) \cos(t - x - \theta - \Psi(x,\theta)) + \zeta(t;x,\theta) \quad ,$$

and

$$A(x,\theta) > 0 \quad .$$

The continuous differentiability of $A(x,\theta)$ follows from that of $B(x,\theta)$ and $C(x,\theta)$. From the definition of $B(x,\theta)$ and $C(x,\theta)$ (Eq. (8.7.9)) it is easy to see that

$$\lim_{x\to\infty} A(x,\theta) = 1 \quad .$$

Moreover, also from that equation we note that for large x (recall we always consider $t \geq x$)

$$u(t; x, \theta) = \cos(t - x - \theta) + \zeta(t; x, \theta) \tag{8.7.18}$$

where

$$\lim_{x \to \infty} \zeta(t; x, \theta) = 0 \quad .$$

Thus we may require

$$\lim_{x \to \infty} \Psi(x, \theta) = 0 \quad .$$

It is now clear that Ψ may also be defined in such a way as to be continuously differentiable.

8.8 Partial Differential Equations For A and Ψ

Having established that A and Ψ have continuous partial derivatives with respect to x and θ we are in a position to find equations for these quantities. Let $\Delta \to 0$ and consider $u(x + \Delta x, \theta)$. From (8.7.5) we compute

$$u(x + \Delta; x, \theta) = \cos\theta + \Delta \sin\theta + o(\Delta) \quad . \tag{8.8.1}$$

Similarly, on the differentiating (8.7.5) with respect to t;

$$u'(x + \Delta; x, \theta) = \sin\theta + \Delta(f(x) - 1)\cos\theta + o(\Delta) \quad . \tag{8.8.2}$$

Now we recognise that these values of u and u' may be considered as initial values for a new problem starting at $x + \Delta$. Thus

$$\begin{aligned} u(x + \Delta; x, \theta) &= \eta(x, \theta)\, u(x + \Delta; x + \Delta, \theta') \\ &= \eta(x, \theta) \cos\theta' \\ u(x + \Delta; x, \theta) &= \eta(x, \theta)\, u'(x + \Delta; x + \Delta, \theta') \\ &= \eta(x, \theta) \sin\theta' \quad . \end{aligned} \tag{8.8.3}$$

Here the factor η is essential since there is no assurance that

$$u^2(x + \Delta; x, \theta) + u'^2(x + \Delta; x, \theta) = 1 \quad .$$

From (8.8.1), (8.8.2), and (8.8.3) we easily compute θ' and η:

$$\theta' = \theta + (f(x)\cos^2\theta - 1)\Delta + o(\Delta) \quad , \tag{8.8.4}$$

$$\eta = 1 + \cos\theta \sin\theta f(x)\Delta + o(\Delta) \quad .$$

Further, for all $t \geq x + \Delta$,

$$u(t; x, \theta) = \eta(x, \theta)\, u(t; x + \Delta, \theta') \quad . \tag{8.8.5}$$

so that

$$\begin{aligned} &A(x,\theta)\cos(t - x - \theta - \Psi(x,\theta)) + \zeta(t; x, \theta) \\ &= \eta(x,\theta)\{A(x+\Delta, \theta')\cos(t - x - \Delta - \theta' - \Psi(x+\Delta, \theta')) \\ &\qquad + \zeta(t; x+\Delta, \theta')\} \end{aligned} \tag{8.8.6}$$

Thus it readily follows that

$$- \theta - \Psi(x, \theta) = -\Delta - \theta' - \Psi(x + \Delta, \theta') \quad . \tag{8.8.7}$$

$$A(x, \theta) = \eta(x, \theta)\, A(x + \Delta, \theta') \quad .$$

Since the partial derivatives of Ψ and A exist and are continuous, the obvious computations with the use of (8.8.4) finally yield

$$\frac{\partial\Psi}{\partial x} + \frac{\partial\Psi}{\partial\theta}(f(x)\cos^2\theta - 1) = -f(x)\cos^2\theta \quad , \tag{8.8.8a}$$

$$\frac{\partial A}{\partial x} + \frac{\partial A}{\partial\theta}(f(x)\cos^2\theta - 1) = -\cos\theta\sin\theta f(x) A \quad . \tag{8.8.8b}$$

8.9 The Solution of (8.8.8)

Equations (8.8.8) must now be solved subject to the conditions of (8.7.14). Concentrating on the equation for Ψ and using the method of characteristics gives

$$(8.9.1)\qquad \frac{dx}{ds} = 1\ , \quad \frac{d\theta}{ds} = f(x)\cos^2\theta - 1\ , \quad \frac{d\Psi}{ds} = -f(x)\cos^2\theta \quad .$$

Now require $s = x$ and $\theta = \theta_o$ when $s_o = x_o$. Then (8.9.1) has the unique solution provided by

$$(8.9.2)\qquad \Psi(x_o, \theta_o) = \int_{x_o}^{\infty} f(s)\cos^2\theta(s)\,ds \quad ,$$

where

$$(8.9.3)\qquad \frac{d\theta}{ds}(s) = f(s)\cos^2\theta(s) - 1 \quad ,$$

with $\theta = \theta_o$ at $s = x_o$. That Ψ satisfies (8.7.14) follows at once from the condition (8.7.1).

Similar arguments with the partial differential equation (8.8.8b) yield

$$(8.9.4)\qquad A(x_o, \theta_o) = \exp\left\{\frac{1}{2}\int_{x_o}^{\infty} f(s)\sin 2\theta(s)\,ds\right\}$$

where $\theta(s)$ is also given by (8.9.3). Notice that

$$\lim_{x_o\to\infty} A(x_o, \theta_o) = 1\ ,$$

as it should.

We state our conclusions formally, changing notation slightly:

THEOREM 2. Let $u(t; x, \theta)$ be the solution to the problem posed in (8.7.2). Then

$$u(t; x, \theta) = A(x, \theta)\cos(t - x - \theta - \Psi(x, \theta)) + \zeta(t; x, \theta)$$

where

$$\lim_{t\to\infty} \zeta(t; x, \theta) = 0$$

uniformly in x and θ, and

$$\Psi(x, \theta) = \int_x^\infty f(s) \cos^2 \theta(s)\, ds \quad ,$$

$$A(x, \theta) = \exp\left\{\frac{1}{2}\int_x^\infty f(s) \sin 2\theta(s)\, ds\right\} \quad ,$$

$\theta(s)$ being given as the solution to the ordinary differential system

$$\frac{d\theta}{ds}(s) = f(s) \cos^2 \theta(s) - 1 \ ,$$

$$\theta(x) = \theta_0 \quad \text{at} \quad s = x_0 \quad .$$

8.10 Connection With The Results Of Franchetti

In 1957, Franchetti, using a method quite different from that just outlined, studied the phase shift problem. A very similar approach has recently been employed by Levy and Keller. These authors studied the problem of (8.7.2) with $x = 0$ and $\theta = \pi/2$. Their solution is written in the form

(8.10.1) $$u(t) = \tilde{A}(t) \sin (t + \tilde{\eta}(t))$$

and an equation for $\tilde{\eta}$ is derived:

(8.10.2) $$\tilde{\eta}'(t) = -f(t) \sin^2 (t + \tilde{\eta}(t)) \quad ,$$

$$\tilde{\eta}(0) = 0 \quad .$$

We wish to show the equivalence of their solutions to ours. Turn to Eq. (8.9.1) and set

(8.10.3) $$\theta(s) + s = + \frac{\pi}{2} - \bar{\eta}(s) \quad .$$

Then

(8.10.4) $$\frac{d\tilde{\eta}}{ds} = -f(s) \cos^2 (\frac{\pi}{2} - \bar{\eta}(s) - s)$$

$$= -f(s) \sin^2 (s + \bar{\eta}(s)) \quad .$$

Also, using $x_o = 0$, $\theta_o = \pi/2$,

$$(8.10.5) \qquad \theta(0) = \frac{\pi}{2} = \frac{\pi}{2} - \bar{\eta}(0)$$

so that

$$(8.10.6) \qquad \bar{\eta}(0) = 0 \quad .$$

Thus $\tilde{\eta}(s)$ satisfies (8.10.2) and so $\tilde{\eta} \equiv \bar{\eta}$. In the treatments the phase shift is given by

$$(8.10.7) \qquad \lim_{t\to\infty} \tilde{\eta}(t) = \bar{\eta}_\infty = \tilde{\eta}_\infty \quad .$$

Hence, writing (8.7.13) with $x = x_o = 0$, $\theta = \theta_o = \pi/2$, we expect

$$(8.10.8) \qquad \lim_{s\to\infty} \bar{\eta}(s) = -\Psi(0, \frac{\pi}{2}) \quad .$$

To see that this is indeed so, note from (8.9.1) that

$$(8.10.9) \qquad \frac{d}{ds}(\theta(s) + \Psi(s)) = -1 \quad ,$$

so that

$$(8.10.10) \qquad \theta(s) + \Psi(s) = -s + C \quad .$$

Using (8.10.3)

$$(8.10.11) \qquad \frac{\pi}{2} - \bar{\eta}(s) + \Psi(s) = C \quad .$$

If we let $s \to \infty$ we get

$$(8.10.12) \qquad \frac{\pi}{2} - \bar{\eta}_\infty = C \quad .$$

Hence

$$(8.10.13) \qquad \theta(s) + \Psi(s) = -s + \frac{\pi}{2} - \bar{\eta}_\infty \quad .$$

Now set $\theta = \pi/2$ when $s = x = 0$ to get

(8.10.14) $\Psi(0, \pi/2) = -\bar{\eta}_\infty$;

which is precisely (8.10.8).

Thus our results include the earlier ones.

8.11 A Representation Of The Solution For Large t

Let $f(t)$ again satisfy the conditions of Sec. 8.7 and consider

(8.11.1a) $$\frac{d^2y}{dt^2} - y = f(t)y , \qquad y(t) = y(t;\, x,\, \theta);\ t \geq x ,$$

(8.11.1b) $$y(x;\, x,\, \theta) = \cos\theta ,$$

(8.11.1c) $$\left.\frac{dy}{dt}\right|_x = y'(x;\, x,\, \theta) = \sin\theta .$$

Here it is known that there are two independent solutions to (8.11.1a):

(8.11.2) $$y_1(t) = e^t(1 + o(1)) ,$$

$$y_2(t) = e^{-t}(1 + o(1)) .$$

We proceed as in Sec. 8.7. Write

(8.11.3) $$y(t;\, x,\, \theta) = \cos\theta \cosh(t-x) + \sin\theta \sinh(t-x) + \int_x^t \sinh(t-w)\, f(w)\, y(w;\, x,\, \theta)dw .$$

It is convenient to introduce

(8.11.4) $$z(t;\, x,\, \theta) = e^{-t+x}\, y(t;\, x,\, \theta) ,$$

so that

$$(8.11.5) \quad z(t;\, x,\, \theta) = e^{-t+x} \cos\theta \cosh(t-x) + e^{-t+x} \sin\theta \sinh(t-x)$$

$$+ \int_x^t e^{-t+x} \sinh(t-w)\, e^{w-x}\, z(w;\, x,\, \theta)\, dw$$

$$= \frac{\cos\theta + \sin\theta}{2} + \frac{\cos\theta - \sin\theta}{2} e^{-2t+2x}$$

$$+ \int_x^t \frac{1 - e^{-2(t-w)}}{2} f(w)\, z(w;\, x,\, \theta)\, dw \quad .$$

Thus, since $t \geq x \geq 0$,

$$(8.11.6) \quad |z(t;\, x,\, \theta)\,| \leq 2 + \frac{1}{2} \int_x^t |f(x)|\, |z(w;\, x,\, \theta)|\, dw \quad .$$

LEMMA 3. The function $z(t;\, x,\, \theta)$ satisfies

$$(8.11.7) \quad |z(t;\, x,\, \theta)| \leq 2 \exp\left\{\frac{1}{2} \int_x^t |f(w)|\, dw\right\} \leq M_2$$

where M_2 is independent of x and t.

PROOF. The lemma is an immediate result of the inequality of Gronwall and the conditions put upon $f(t)$.

We may now write $z(t;\, x,\, \theta)$ in another form:

$$(8.11.8) \quad z(t;\, x,\, \theta) = \left\{\frac{\cos\theta + \sin\theta}{2} + \frac{1}{2} \int_x^\infty f(w)\, z(w;\, x,\, \theta)\, dw\right\}$$

$$+ \frac{\cos\theta - \sin\theta}{2} e^{-2t+2x} - \frac{1}{2} \int_x^\infty f(w)\, z(w;\, x,\, \theta)\, dw$$

$$+ \int_x^t e^{-2(t-w)} f(w)\, z(w;\, x,\, \theta)\, dw$$

$$= A(x,\, \theta) + \xi(t;\, x,\, \theta) \quad .$$

where

$$(8.11.9) \qquad A(x, \theta) = \frac{\cos\theta + \sin\theta}{2} + \frac{1}{2}\int_x^\infty f(w)\, z(w; x, \theta)\, dw \quad .$$

LEMMA 4. $A(x,\theta)$ has continuous first partial derivatives with respect to both x and θ. Moreover

$$(8.11.10) \qquad \lim_{t\to\infty} \xi(t; x, \theta) = 0 \quad .$$

PROOF. The first part of the lemma is proved exactly as is the first part of Lemma 2. We leave the details to the reader.

The remaining portion of the lemma is also easy except perhaps for the behavior of one term. Write

$$(8.11.11) \qquad I = \int_x^t e^{-2(t-w)} f(w)\, z(w; x, \theta)\, dw$$

$$= e^{-2t}\int_x^{t/2} e^{2w} f(w)\, z(w; x, \theta)\, dw$$

$$+ e^{-2t}\int_{t/2}^{t} e^{2w} f(w)\, z(w; x, \theta)\, dw \quad .$$

Thus

$$(8.11.12) \qquad |I| \le e^{-2t}e^{t} M_2 \int_x^{t/2} |f(w)|\, dw + M_2 \int_{t/2}^{t} |f(w)|\, dw \quad .$$

Since

$$\lim_{t\to\infty} \int_{t/2}^{t} |f(w)|\, dw = 0$$

the result on ξ follows.

THEOREM 3. The solution $y(t; x, \theta)$ of (8.11.1) may be written in the form

$$y(t; x, \theta) = e^{t-x}\left\{A(x, \theta) + \xi(t; x, \theta)\right\}$$

where $A(x, \theta)$ has continuous first partial derivatives with respect to x and θ,

$$(8.11.13) \qquad \lim_{x \to \infty} A(x, \theta) = \frac{\cos\theta + \sin\theta}{2}$$

and

$$\lim_{t \to \infty} \xi(t; x, \theta) = 0 \quad .$$

PROOF. The results follow from the foregoing lemmas and from (8.11.9). The condition (8.11.13) is not a convenient one to deal with. We therefore define

$$(8.11.14) \qquad B(x, \theta) = A(x, \theta) - \frac{\cos\theta + \sin\theta}{2} \quad ,$$

so that

$$(8.11.15) \qquad y(t; x, \theta) = e^{t-x} \left\{ \left[B(x, \theta) + \frac{\cos\theta + \sin\theta}{2} \right] + \xi(t; x, \theta) \right\}$$

and

$$(8.11.16) \qquad \lim_{x \to \infty} B(x, \theta) = 0 \quad .$$

8.12 A Partial Differential Equation For $B(x, \theta)$ And Its Solution

The results of Sec. 8.11 now allow us to find a partial differential equation satisfied by $B(x,\theta)$. The technique is exactly the same as that used in Sec. 8.8, although the computations are somewhat messier. The final result, which we merely state, is

$$(8.12.1) \qquad \frac{\partial B}{\partial x} + (\cos 2\theta + f(x) \cos^2\theta) \frac{\partial B}{\partial \theta} = B \cdot \left[1 - \sin 2\theta \left(1 + \frac{f(x)}{2} \right) \right] - \frac{f(x)}{2} \cos\theta \quad ,$$

$$\lim_{x \to \infty} B(x, \theta) = 0 \quad .$$

The method of characteristics yields the equations for B.

THEOREM 4. The function $B(x, \theta)$ is given by

$$B(x, \theta) = 1/2 \int_x^\infty f(w) \cos \theta(w) \exp \left\{ - \int_x^w \left[1 - \sin 2\,\theta(p) \left(1 + \frac{f(p)}{2}\right)\right] dp \right\} dw \quad , \tag{8.12.2}$$

with

$$\frac{d\theta(s)}{ds} = \cos 2\,\theta(s) + f(s) \cos^2 \theta(s) \quad ,$$

$$\theta(x) = \theta_o \quad .$$

We note that the integral giving B is convergent and that B has the desired bebavior as $x \to \infty$.

8.13 Some Remarks On More General Cases

We have already mentioned in the introduction that the ideas introduced here seem capable of considerable generalization. Thus, suppose we wish to study the asymptotic behavior of the solution to

$$Lu = f(t)\, u \quad , \tag{8.13.1}$$

where L is a quite general linear operator, and enough initial conditions are imposed upon u to ensure a unique solution to the problem. Let the fundamental solutions to

$$Lu = 0 \tag{8.13.2}$$

be known explicitly. If $f(t)$ is in some sense "small" for large t the techniques employed can then be formally applied. The ultimate success of the scheme depends, of course, upon the detailed properties of L and f. A study of Secs. 8.7 through 8.12 reveals that there are likely to be only a few "sticky" points.

As an example, we cite

$$\frac{d^2u}{dt^2} + \left(k^2 - \frac{I(I+1)}{t^2} \right) u = V(t)\, u(t) \quad , \tag{8.13.3}$$

an equation of great interest in quantum mechanical scattering theory. Since the fundamental solutions to (8.13.3) when $V \equiv 0$ are well known, our general prescription would seem applicable. Details have not yet been carried out, however.

8.14 The Poincaré - Lyapunov Theorem

Consider the nonlinear vector differential equation

$$x' = Ax + g(x) \;, \qquad x(0) = c \quad . \tag{8.14.1}$$

If we assume

(8.14.2)
1. All the characteristic roots of A have negative real parts; i.e., A is a stability matrix.
2. $\|c\|$ is sufficiently small.
3. $g(x)$ is a power series in the components of x lacking constant and first degree terms,

then we may conclude that the solution exists for all t. Furthermore,

$$x \sim e^{\lambda t} f(c) \quad . \tag{8.14.3}$$

Here λ is the characteristic root of A with largest real part. Let us assume for the moment that λ is real and simple.

The purpose of these sections is to determine the function $f(c)$. As we shall see we can use a simple invariant imbedding argument.

Then we turn to the question of the asymptotic behavior of

the other solutions. This is an interesting question. To simplify the algebra, we shall consider the equation

$$u'' - 3u' + 2u = u^2 \quad . \tag{8.14.4}$$

Next, we consider the case where some of the characteristic roots have positive real part. Again to simplify the algebra we shall consider the equation,

$$u'' - u = u^2 \quad . \tag{8.14.5}$$

Then we shall consider the case where some of the characteristic roots have zero real part. We shall consider the equation

$$u'' + u + f(t)\, g(u) = 0 \quad , \tag{8.14.6}$$

where we assume $\int_0^\infty |f|dt < \infty$.

Finally, we shall show that $f(c)$ has a very interesting linearizing property.

8.15 The Determination of $f(c)$

One way to establish the theorem of Poincaré and Lyapunov is to consider the integral equation

$$x = e^{At} c + \int_0^t e^{A(t-t_1)} g(x)\, dt_1 \quad . \tag{8.15.1}$$

We now employ the method of successive approximation. The function $f(c)$ may not be found by iteration. We shall follow a different and easier approach.

We have

$$\lim_{t\to\infty} x(t)\, e^{-\lambda t} = f(c) \quad . \tag{8.15.2}$$

This may also be written

$$\lim_{t \to \infty} x(t + \Delta)\, e^{-\lambda(t + \Delta)} = f(c) \quad . \tag{8.15.3}$$

Let us now regard Δ as an infinitesimal. Then we have, to terms in Δ^2,

$$e^{-\lambda\Delta} f(c + \Delta\,(Ac + g(c))) = f(c) \quad . \tag{8.15.4}$$

Passing to the limit as Δ approaches zero, we obtain a linear partial differential equation for $f(c)$. From this equation, we can usually determine the coefficients of $f(c)$ by recurrence.

8.16 The Asymptotic Behavior Of Other Solutions

In the foregoing section, we determined the asymptotic behavior of a general solution. It is interesting to ask for the asymptotic behavior of particular solutions. This is of interest both analytically and computationally. If we try to find a solution which has small asymptotic behavior by computation, then round-off error may bring in the larger solution. Here, there are two problems. We have to determine the asymptotic behavior and suitable initial values.

As we know, we have the integral equation

$$u = c_1 e^{-t} + c_2 e^{-2t} + \int_0^t (e^{-2(t-2)} - e^{-(t-s)})\, u^2(s)ds \quad . \tag{8.16.1}$$

The value of c_1 must now be chosen so as to eliminate the term e^{-t}. Let us consider then the integral equation

$$u = c_2 e^{-2t} + e^{-t} \int_t^\infty e^s u^2(s)ds + e^{-2t} \int_0^t e^{2s} u^2(s)ds \quad . \tag{8.16.2}$$

The solution of this integral equation may be found by successive approximation.

We see that to determine the asymptotic behavior we require the integral $\int_0^\infty e^{2s}u^2 ds$; to determine the initial values we require the integral $\int_0^\infty e^s u^2 ds$. The initial values and its derivative are simple linear functions of this integral.

Let us set

$$(8.16.3) \qquad u(0) = f(c_2) , \qquad u'(0) = g(c_2)$$

$$h(f(c_2)) = \int_0^\infty e^s u^2 ds \quad .$$

We see that h could be written as a function of c_2. The representation we use here makes the derivation of the differential equation satisfied by h simpler.

We will evaluate both integrals using the same method. Hence, it is sufficient to consider the first. We have

$$(8.16.4) \qquad \int_0^\infty = \int_0^\Delta + \int_\Delta^\infty \quad .$$

where we assume that Δ is an infinitesimal. In the second integral, we perform a change of variable so that the lower limit is zero. We then have

$$(8.16.5) \qquad h(f(c_2)) = \Delta f(c_2)^2 + e^{2\Delta} h(f(c_2) + \Delta g(c_2)) + \ldots$$

to terms in Δ^2. Passing to the limit, we obtain a differential equation for $h(f(c_2))$. The coefficients can now easily be determined recurrently. The other integral may be evaluated in the same fashion.

8.17 Some Characteristic Roots With Positive Real Part

The same procedure may be used to treat the case where some of the characterisitic roots have positive real part. To simplify the algebra, we can apply the method to Eq. (8.14.5). The integral equation we use is very much as above and the method we employ is the same.

8.18 Some Characteristic Roots with Zero Real Part

Let us now consider Eq. (8.14.6). We use the integral equation as above. We now follow the method given in (8.14.1) to find the desired asymptotic behavior.

8.19 An Important Functional Equation

Let us now return to Sec. 8.2. We have

$$\lim_{t\to\infty} x(t)\, e^{-\lambda t} = f(c) \quad . \tag{8.19.1}$$

This may also be written

$$\lim_{t\to\infty} x(t+1)\, c^{-\lambda(t+1)} = f(c) \quad . \tag{8.19.2}$$

Let the solution at $t=1$ be denoted by $h(c)$. Then we have

$$f(h(c)) = e^{\lambda} f(c) \quad . \tag{8.19.3}$$

We see then that with the aid of f the iterates of h may be obtained easily by multiplication.

Bibliography and Comments

Section 8.1. For the theory of invariant imbedding, see Bellman, R., and Wing, G. M., An Introduction to Invariant Imbedding, John Wiley & Sons, Inc., New York, 1974.

Section 8.6. We are following the paper, Wing, G. M., "Invariant Imbedding and the Asymptotic Behavior Solutions to Initial Value Problems", Journal of Mathematical Analysis and Applications, Vol. 9, No. 1, August 1964, pp. 85-98.

For a multidimensional version, see Bellman, R., "Invariant Imbedding and Asymptotic Behavior", Journal of Mathematical Analysis and Applications, Vol. 22, 1968, pp. 444-447.

Section 14. We are following the paper, Bellman, R., "On the Poincaré-Lyapunov Theorem", Journal of Nonlinear Analysis: Theory, Methods and Applications, Vol. 4, No. 2, 1980, pp. 297-300.

Chapter IX

PARTIAL DIFFERENTIAL EQUATIONS AND DYNAMIC PROGRAMMING

9.1 Introduction

In this chapter, we want to describe some of the connections between the calculus of variations and partial differential equations using dynamic programming.

We can then proceed in either direction. We can solve the variational problem using the partial differential equation. This procedure has several advantages. First of all it is an initial value method. Secondly, constraints which make a classical approach so difficult simplify the dynamic programming approach. Thirdly, the dynamic programming approach suggests new methods for treating the partial differential equation. Fourthly, we can obtain various bounds for the partial differential equation in this way. We say a few words about that below.

We can also proceed in the other direction. In many cases we don't want to solve the partial differential equation for all values of the variables. In that case, we can solve the associated variational problem for these values. In order to solve the variational problem, in general we have to employ some method of approximation. Some of these are described in the references at the end of the chapter.

We show that more complicated partial differential equations can be obtained by considering more complicated variational problems involving layered functionals.

9.2 Calculus of Variations as a Multistage Decision Process

It is of considerable interest to examine the fiction of continuous decision processes. In particular, we wish to demonstrate that we can profitably regard the calculus of variations as an example of a multistage decision process of continuous type. In consequence of this, dynamic programming provides a number of new conceptual, analytic, and computational approaches to classical and modern variational problems, particularly to those arising in control processes.

To illustrate the basic idea, which is both quite simple and natural from the standpoint of a control process, let us consider the scalar functional

$$J(u) = \int_0^T g(u, u')\, dt \quad . \tag{9.2.1}$$

The problem of minimizing $J(u)$ over all u satisfying the initial condition $u(0) = c$ leads along classical lines as we know to the task of solving the Euler-Lagrange equation,

$$\frac{\partial g}{\partial u} - \frac{d}{dt}\left(\frac{\partial g}{\partial u'}\right) = 0 \quad , \tag{9.2.2}$$

subject to the two-point condition

$$u(0) = c \ , \qquad \left.\frac{\partial g}{\partial u'}\right|_{t-T} = 0 \quad . \tag{9.2.3}$$

This equation, is a variational equation obtained by considering the behavior of the functional $J(u+w)$ for "all" small w where u is the desired minimizing function. This procedure of examining the neighborhood of the extremal in function space is a natural generalization of that used in calculus in the finite-dimensional case. As in the finite-dimensional case, there can be considerable difficulty first in solving the variational equation and then in distinguishing the absolute minimum from other stationary points.

Let us now pursue an entirely different approach motivated by the theory of dynamic programming. In particular, it is suggested by the applications of dynamic programming to the study of deterministic control processes. In place of thinking of a curve $u(t)$ as a locus of points, let us take it to be an envelope of tangents; see Figs. 8.1 and 8.2. Ordinarily, we determine a point on the curve by the coordinates $(t, u(t))$. However, we can equally well trace out the curve by providing a rule for determining the slope u' at each point (t, u) along the path. The determination of the minimizing curve $u(t)$ can thus be regarded as a multistage decision process in which it is necessary to choose a direction at each point along the path. Motivation of this approach in the domain of pursuit processes is easily seen, or equivalently, in the determination of geodesics, in the analysis of multistage investment processes, or in the study of optimal growth processes in mathematical economics.

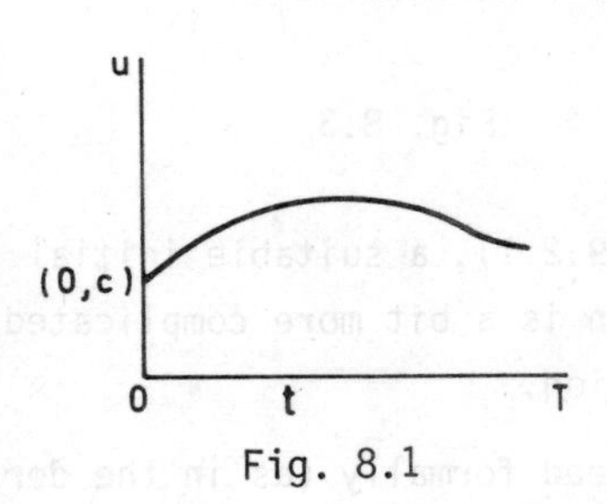

Fig. 8.1

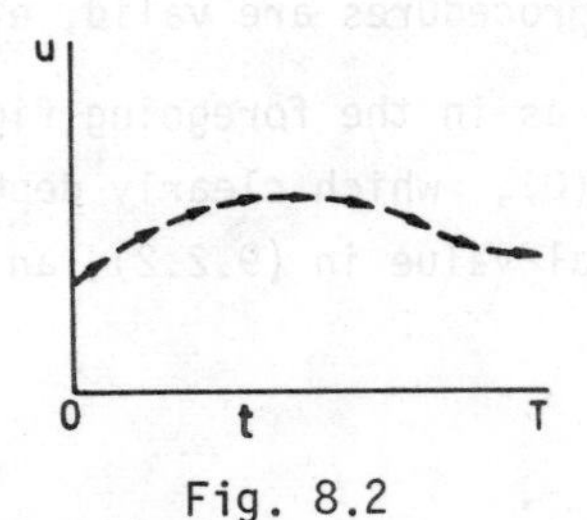

Fig. 8.2

9.3 A New Formalism

Let us use the foregoing concept of a minimization process to obtain a new analytic approach to variational problems. See Fig. 8.3. Let us denote the minimum value of $J(u)$ as defined by (9.2.1) (assumed to exist) by $f(c, T)$. Thus, we introduce the function

$$f(c, T) = \min_u J(u) , \tag{9.3.1}$$

defined for $T \geq 0$ and $-_\infty < c < \infty$. Since we are now interested in policies, the initial state must be considered to be a variable.

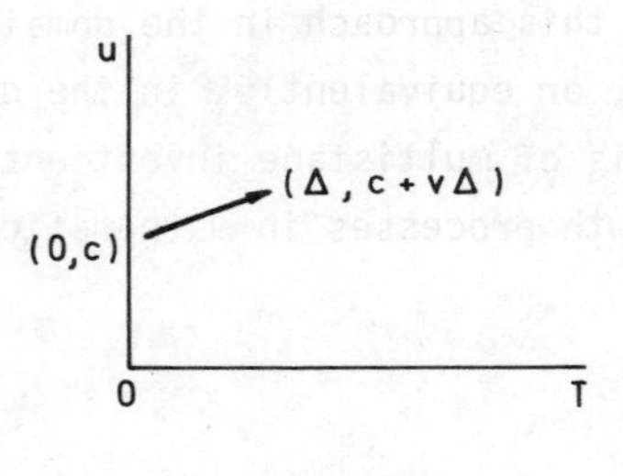

Fig. 8.3

For the problem (9.2.1), a suitable initial condition is $f(c, 0) = 0$. The situation is a bit more complicated when there is a two-point boundary condition.

Let us first proceed formally (as in the derivation of the Euler-Lagrange equation), supposing that all partial derivatives exist, that all limiting procedures are valid, etc.

Let $v = v(c, T)$, as in the foregoing figure, denote the initial direction; $v = u'(0)$, which clearly depends on c and T. This is the missing initial value in (9.2.2), an essential quantity. Writing

$$\int_0^T = \int_0^\Delta + \int_\Delta^T , \tag{9.3.2}$$

convenient separability property of the integral, we see that for ny initial $v(c, T)$ we have

$$f(c, T) = \Delta g\ (c, v) + O(\Delta^2) + \int_\Delta^T \qquad (9.3.3)$$

ere Δ is an infinitesimal. We now argue as follows. Regardless f how $v(c, T)$ was chosen, we are going to proceed from the new oint $(\Delta, c+v\Delta)$ so as to minimize the remaining integral $\int_\Delta^T g(u,u')dt$. ut

$$\min_u \int_\Delta^T g(u, u')dt = f(c + v\Delta, T - \Delta) \qquad (9.3.4)$$

y definition of the function f. Hence, for any initial choice of , we have

$$f(c, T) = g(c, v)\ \Delta + f(c + v\Delta, T - \Delta) + O(\Delta^2) \qquad (9.3.5)$$

his is an example of the "principle of optimality" for multistage ecision processes.

It remains to choose $v(c, T)$ appropriately, Clearly, v hould be chosen to minimize the right-hand side of (9.3.5). Thus, e obtain the equation

$$f(c, T) = \min_v \left[g(c, v)\ \Delta + f(c + v\Delta, T - \Delta)\right] + O(\Delta^2) \qquad (9.3.6)$$

xpanding the appropriate terms above in powers of Δ and letting $\to 0$, we obtain the partial differential equation

$$\frac{\partial f}{\partial T} = \min_v \left[\ g(c, v) + \frac{\partial f}{\partial c}\ \right] \qquad (9.3.7)$$

e have noted above that $f(c, 0) = 0$, an initial condition. Thus e have transformed the original variational problem of minimizing (u) in (9.2.1) into that of solving a nonlinear partial differential equation subject to an initial condition.

The foregoing is a cavalier approach in the spirit of the usual first derivation of the Euler-Lagrange equation. We leave it to the reader to spot all the irregularities.

9.4 Riccati Equations

If the functional is quadratic, the partial differential equation reduces to an ordinary differential equation. In the one-dimensional case, we have a separation of variables

$$f(c, t) = c^2 r(t) \quad . \tag{9.4.1}$$

The function $r(t)$ satisfies a Riccati equation. In the multidimensional case, we obtain in this fashion a matrix Riccati equation.

Since the functional is quadratic, we can apply the Rayleigh-Ritz method easily. The variational equations are now linear algebrai equations.

9.5 Layered Functionals

Let us call a functional of the form

$$J(u) = \int_0^T g\left(u, u', \int_0^T h(u, u')dt_1\right) dt \tag{9.5.1}$$

a layered functional. The problem of minimizing $J(u)$ with respect to u, subject to an initial condition $u(0) = c$, can be approached in the following fashion. Set

$$\int_0^T h(u, u')dt = k \quad , \tag{9.5.2}$$

where k is a parameter to be determined, and consider the more usual problem of minimizing the functional

$$J(u, k) = \int_0^T g(u, u', k)dt \quad . \tag{9.5.3}$$

This determines a function $u(t, k)$. The value of k is to be obtained from the consistency relation

$$\int_0^T h(u(t, k), u'(t, k))dt = k \quad . \tag{9.5.4}$$

Clearly, there are some serious analytical and computational obstacles in the path of an approach of this nature.

Let us then examine an alternative approach using dynamic programming. In this fashion we are led to some initial-value problems for nonlinear partial differential equations. Conversely, we are led to some representation theorems for certain classes of nonlinear partial differential equation. These representation theorems can be used to obtain upper, and in some cases lower, bounds for the solutions. In what follows we will present the purely formal aspects.

9.6 Dynamic Programming Approach

Let us take the more general problem of minimizing

$$J(u, a) = \int_0^T g\left(u, u', a + \int_0^T h(u, u')dt\right)dt \tag{9.6.1}$$

subject to $u(0) = c$, where a is a parameter in the range $(-\infty, \infty)$. Introduce the function

$$f(c, a, T) = \min_u J(u, a) \quad , \tag{9.6.2}$$

and write

$$\phi(c, a, T) = \int_0^T h(u, u')dt \quad , \tag{9.6.3}$$

where u is the function, assumed to exist, which minimizes $J(u, a)$.

Then, proceeding in a familiar fashion, we write

$$(9.6.4) \qquad f(c, a, T) = \min_u \left[\int_0^\Delta + \int_\Delta^T \right] ,$$

leading via the principle of optimality to

$$(9.6.5) \qquad f(c, a, T) = \min_v \Big[g(c, v, a + \phi)\Delta$$

$$+ f(c + v\Delta, a + h(c, v)\Delta, T - \Delta) \Big] + O(\Delta) ,$$

and thus, in the limit as $\Delta \to 0$, to the partial differential equation

$$(9.6.6) \qquad \frac{\partial f}{\partial T} = \min_v \left[g(c, v, a + \phi) + v \frac{\partial f}{\partial c} + h(c, v) \frac{\partial f}{\partial a} \right] .$$

Similarly, ϕ satisfies the equation

$$(9.6.7) \qquad \frac{\partial \phi}{\partial T} = (hc, v) + v \frac{\partial \phi}{\partial c} + h(c, v) \frac{\partial \phi}{\partial a} ,$$

where $v = v(\phi, \frac{\partial f}{\partial c}, \frac{\partial f}{\partial a}, \frac{\partial f}{\partial T})$ is determined by (9.6.6). The initial conditions are

$$(9.6.8) \qquad f(c, a, 0) = 0 , \qquad \phi(c, a, 0) = 0 .$$

We can obtain more general boundary conditions by taking the functional

$$(9.6.9) \qquad J_1(u, a) = \int_0^T g \left(u, u', a + r(u(T)) + \int_0^T h(u, u')dt \right) d$$

$$+ s(u(T))$$

as a starting point. Then (9.6.8) is replaced by

$$(9.6.10) \qquad f(c, a, 0) = s(c), \qquad \phi(c, a, 0) = r(c) .$$

9.7 Quadratic Case

We can obtain particular classes of quadratically nonlinear partial differential equations in this way by choosing various quadratic functionals such as

$$J_2(u, a) = \int_0^T \left(a + u' + \int_0^T u \, dt\right)^2 dt + \int_0^T u^2 \, dt \quad . \tag{9.7.1}$$

Furthermore, the functional

$$J_3(u, a) = \int_0^T \left[u'^2 + g\left(a + u + \int_0^T u \, dt \right) \right] dt \tag{9.7.2}$$

leads to an interesting type of nonlinear partial differential equation.

9.8 Bounds

Since we have a variational problem where we are seeking the minimum, we know that any trial function provides an upper bound. In this way, we readily obtain upper bounds for the solution of the associated partial differential equation.

To obtain lower bounds, we can proceed in various fashions. First of all, we can use a dual problem, a procedure introduced by Friedrichs, to obtain lower bounds. Secondly, we can find another problem for the desired quantity which yields a lower bound. Thirdly, we can use geometrical ideas to obtain lower bounds.

Bibliography and Comments

Section 9.1. A description of dynamic programming and variational problems may be found in Bellman, R., Introduction to the Mathematical Theory of Control Processes, Vol. II, Academic Press, Inc., New York, 1971.

Quadratic functionals are discussed in detail in, Bellman, R., Introduction to the Mathematical Theory of Control Processes, Vol. I, Academic Press, Inc., New York, 1968.

Approximation methods are given in the first book cited above and in Bellman, R., Methods of Nonlinear Analysis, Vol. I, Academic Press, Inc., New York, 1969.

Section 9.5. We are following the paper, Bellman, R., "Functional Equations in the Theory of Dynamic Programming. XV. Layered Functionals and Partial Differential Equations", Journal of Mathematical Analysis and Applications, Vol. 28, No. 1, October 1969, pp. 1-3.

Section 9.8. Geometric methods are discussed in Beckenbach, E. F., and Bellman, R., Inequalities, Springer-Verlag, Berlin, 1970.

Rockefeller, T., Convex Analysis, Princeton University Press, Princeton, New Jersey, 1970.

Chapter X

PARTIAL DIFFERENTIAL EQUATIONS AND INVARIANT IMBEDDING

10.1 Introduction

In this chapter, we want to give some of the connections between invariant imbedding and partial differential equations.

First, we show how a two-point boundary value problem leads to nonlinear partial differential equations. As in the previous chapter, we can now go in either direction. We can solve the two-point boundary value problem by means of the partial differential equation which is of initial value type, or, if only a few values of the partial differential equation are desired, we can solve the two-point boundary value problems, using various approximate techniques.

Then we show that differential equations subject to integral conditions also lead to nonlinear partial differential equations using invariant imbedding. We point out how, in many cases, this leads to the problem of solving a system of simultaneous equations, and that this technique can be used as an approximation method.

10.2 On The Fundamental Equations of Invariant Imbedding

In 1942, V. Ambarzumian showed how the energy diffusely reflected from a plane-parallel atmosphere can be calculated directly, without first determining the internal flux. This was done through the use of invariance principles, which lead to nonlinear equations, as opposed to the usual linear transport equations for the internal fluxes. These techniques have been greatly extended, but, as I. Busbridge points out in her book, "The application of these principles is not easy and until a precise statement is given of the physical conditions which are sufficient to ensure their truth, any solution based on them ought to be verified in another way." The purpose of this section is to lay bare, in rigorous fashion, the connection between the transport equations for internal fluxes and the invariant imbedding equations for reflected fluxes for a one-dimensional transport process. Linearity of perturbation equations and the uniqueness of solution of a linear two-point boundary value problem are the essential ingredients.

The basic equations for a one-dimensional transport process involving N different types of particles are of the form

$$\frac{dy}{dz} = g(y) , \qquad 0 \le z \le x \quad , \tag{10.2.1}$$

with conditions on the vector at the two points $z = 0$ and x. On the other hand, the invariant imbedding approach leads to quasilinear partial differential equations of the form

$$\frac{\partial w}{\partial x} = \sum_{j=1}^{N} g_j(w) \frac{\partial w}{\partial c_j} + h(w) , \tag{10.2.2}$$

with initial conditions at $x = 0$. These equations, obtained by particle counting directly from the original physical process in the general case, have been derived by Wing in a purely analytical manner in the important case where the vector function $g(y)$ is linear.

We wish to show how to derive equations of type (10.2.2) directly from equations of type (10.2.1) without reference to the underlying physical process.

0.3 The Case $N = 2$

It will be sufficient to discuss the two-dimensional case to llustrate the general method. As we shall see, the derivation of 10.2.2) is a simple consequence of the uniqueness theorem for linear rdinary differential equations

Let u, v be the solutions of

$$\frac{du}{dz} = g(u, v), \qquad u(0) = 0, \tag{10.3.1}$$

$$\frac{dv}{dz} = h(u, v). \qquad v(x) = c, \qquad 0 < z < x,$$

nd let U, V be the solutions of

$$\frac{dU}{dz} = g(U, V), \qquad U(0) = 0, \tag{10.3.2}$$

$$\frac{dV}{dz} = h(U, V), \qquad V(x + \Delta) = c, \qquad 0 < z < x + \Delta,$$

here Δ is an infinitesimal.

The equations (10.3.2) can be converted to the equations with he new boundary condition

$$V(x) = c - V'(x)\Delta = c - h(U(x), V(x))\,\Delta \tag{10.3.3}$$

$$= c - h(u(x), v(x))\,\Delta,$$

terms in $O(\Delta)$. Setting, to terms in $O(\Delta)$,

$$U = u + w\Delta, \qquad V = v + q\Delta, \tag{10.3.4}$$

e see that w and a satisfy the linear perturbation equations

$$\frac{dw}{dz} = wg_u + qg_v, \qquad w(0) = 0, \tag{10.3.5}$$

$$\frac{dg}{dz} = wh_u + qh_v, \qquad q(x) = -h(u(x), v(x)) \quad .$$

Returning to (10.3.1), we see that u_c and v_c satisfy the equations

$$(10.3.6)\qquad \frac{d}{dz}(u_c) = u_c\, g_u + v_c\, g_v \qquad u_c(0) = 0\ ,$$

$$\frac{d}{dz}(v_c) = u_c\, h_u + v_c\, h_v\ , \qquad v_c(x) = 1\ .$$

Assuming that this linear system possesses a unique solution, as is readily established rigorously for small x, we see that

$$(10.3.7)\qquad w = -h(u, v)u_c\ ,$$

$$q = -h(u, v)v_c\ .$$

Since

$$(10.3.8)\qquad U(x + \Delta) = U(x) + \Delta U'(x) = U(x) + \Delta g(u, v)$$

$$= (u(x) + \Delta w) + \Delta g(u, v)\ ,$$

setting $u(x) = r(x, c)$, we see that r satisfies the equation

$$(10.3.9)\qquad \frac{\partial r}{\partial x} = -h(r, c)\,\frac{\partial r}{\partial c} + g(r, c)\ ,$$

an equation for reflected flux derived by particle-counting technique

In addition, the second of equations (10.3.7) and equation (10.3.4) imply that

$$(10.3.10)\qquad \frac{V(0) - v(0)}{\Delta} = -h(u(x), v(x))\, v_c\Big|_{x=0}\ .$$

If we set

$$(10.3.11)\qquad t(x, c) = v(0)\ ,$$

then,

$$\frac{\partial t}{\partial x} = -h(r, c)\frac{\partial t}{\partial c} , \qquad (10.3.12)$$

which is the equation previously found for the transmitted flux.

10.4 Integral Conditions

In this section we will consider the equation

$$u' = f_1(u, v) , \qquad (10.4.1)$$

$$v' = f_2(u, v)$$

subject to the conditions

$$\int_0^T q_1(u, v)\, dt + \phi_1(u(T), v(T)) = k_1 . \qquad (10.4.2)$$

$$\int_0^T q_2(u, v)\, dt + \phi_2(u(T), v(T)) = k_2 .$$

The same method can be applied to higher order equations. However, the second-order case will illustrate the method. We show that a simple invariant imbedding argument yields partial differential equations for the missing initial conditions. We also consider the mixed case where one initial value is given and there is one integral relation.

We shall use a simple technique. We shall calculate the values at Δ in two ways and set them equal. This will yield the required nonlinear partial differential equations.

Let us set

$$u(0) = f_1 (k_1, k_2, T) . \qquad (10.4.3)$$

Then we have, on one hand, to terms in Δ^2,

(10.4.4) $u(\Delta) = f_1 + \Delta g_1 (f_1, f_2)$

and, on the other hand, to terms in Δ^2, we have

(10.4.5) $u(\Delta) = f_1(k_1 - \Delta h_1 (f_1, f_2) k_2 - \Delta h_2 (f_1, f_2) T - \Delta)$.

Let us now equate these values. Passing to the limit as $\Delta \to 0$, we have the required nonlinear partial differential equation.

A similar argument gives the equation for f_2.

10.5 Mixed Conditions

Let us assume that one initial condition is given and that we have one integral relation. Let us set the missing initial condition as $f(c_1, k_1, T)$. The same argument as above now yields a partial differential equation for f.

In many cases it is either not possible to measure the state of a system at a given time or to determine the state accurately. For example, in many medical situations it is impossible to determine the concentration of a drug at a given time; in astronomy it is impossible to measure the light coming from a distant star at a given time. However, if we take a suitable time average, we can determine these quantities quite accurately.

10.6 Selective Computation

If we have a nonlinear partial differential equation of the foregoing type, it may be desired to determine only a few values. In that case, we can solve the foregoing problem. If the equations and the relations are nonlinear, we must use successive approximations.

If the equations are linear, the problem that results is an algebraic one.

Bibliography and Comments

Section 10.1. For the theory of invariant imbedding, see Bellman, R., and Wing, G. M., An Introduction to Invariant Imbedding, John Wiley & Sons, Inc., New York, 1974.

Section 10.4. We are following, Bellman, R., "Selective Computation - IX: Differential Equations Subject to Integral Conditions", Journal of Nonlinear Analysis: Theory, Methods, and Applications, Vol. 4, No. 2, 1980, pp. 301-302.

Chapter XI
MAXIMUM ALTITUDE

11.1 Introduction

The current interest in rockets and space travel has aroused a corresponding interest in the determination of maximum range, minimum time, and so on, for various types of trajectories.

A variety of questions of this type have been treated by means of the theory of dynamic programming. Here we wish to show how to use functional equations to determine the range, the maximum elevations, and similar quantities, as functions of initial position and velocities.

11.2 Vertical Motion - I

Consider an object, subject only to the force of gravity and the resistance of the air, which is propelled straight up. In order to illustrate the technique we shall employ, let us treat the problem of determining the maximum altitude.

Let the defining equation be

$$(11.2.1) \qquad u'' = -g - h(u') ,$$

with the initial conditions $u(0) = 0$, $u'(0) = v$. Here $v > 0$, and $h(u') \geq 0$ for all u'.

Since the maximum altitude is a function of v, let us introduce the function

(11.2.2) $\quad f(v) =$ the maximum altitude attained starting with initial velocity v.

From the definition of the function it follows that

$$f(v) = v\Delta + f(v - [g + h(v)]\Delta) + o(\Delta) \ , \tag{11.2.3}$$

for Δ an infinitesimal. Verbally, this states that the maximum altitude is the altitude gained over an initial time Δ, plus the maximum altitude attained starting with a velocity $v - [g + h(v)]\Delta$, the velocity of the object at the end of time Δ, to within $o(\Delta)$.

Expanding both sides and letting $\Delta \to 0$, we see that

$$f'(v) = \frac{v}{g + h(v)} \quad . \tag{11.2.4}$$

Since $f(0) = 0$, this yields

$$f(v) = \int_0^v \frac{v_1 dv_1}{g + h(v_1)} \quad . \tag{11.2.5}$$

In the particular case where $h(v) = 0$, we obtain the standard result $v^2/2g$.

11.3 Vertical Motion - II

Consider the more general case where motion is through an inhomogeneous medium. Let the defining equation be

$$u'' = h(u, u') \ , \qquad u(0) = c_1 \ , \qquad u'(0) = c_2 \quad . \tag{11.3.1}$$

Assume that $h(u, u') \le 0$ for all u and u', so that $c_2 = 0$ implies no motion.

The maximum altitude is now a function of both c_1 and c_2.

Introduce

(11.3.2) $f(c_1, c_2)$ = The maximum altitude attained starting with the initial position c_1 and initial velocity c_2.

Then, as above,

$$(11.3.3) \qquad f(c_1, c_2) = c_2\Delta + f\left[c_1 + c_2\Delta,\ c_2 + h(c_1, c_2)\Delta\right] + O(\Delta) \ ,$$

which yields in the limit the partial differential equation

$$(11.3.4) \qquad c_2 + c_2 \frac{\partial f}{\partial c_1} + h(c_1, c_2) \frac{\partial f}{\partial c_2} = 0 \quad .$$

By virtue of our assumptions, $f(c_1, 0) \equiv 0$, for $c_1 \geq 0$.

11.4 Computational Aspects

One can, of course, use the method of characteristics, or standard difference methods, to solve (11.3.4). Let us present another method which reduces the solution to the tabulation of a sequence of functions of one variable.

In place of (11.3.4), let us use the discrete approximation of (11.3.3),

$$(11.4.1) \qquad f(c_1, c_2) = c_2\Delta + f\left[c_1 + c_2\Delta,\ c_2 + h(c_1, c_2)\Delta\right] \quad .$$

Since c_2 is monotone decreasing, it can be used to play the role to time. Let us write $c_2 = N\delta$, where δ is a positive quantity, and $f(c_1, c_2) \equiv f_N(c_1)$. We consider then only values of c_2 which are multiples of δ. To overcome the fact that $c_2 + h(c_1, c_2)\Delta$ in general will not be a multiple of δ, we can either replace it by $[(c_2 + h(c_1, c_2)\Delta)/\delta]$ or use interpolation. Although use of an interpolation formula slows up the computation, it greatly improves the accuracy.

11.5 Maximum Altitude

Consider now the case where motion takes place in a plane. Let the equations be

$$(11.5.1) \qquad x'' = g(x', y'), \quad x(0) = 0, \quad x'(0) = c_1,$$

$$y'' = h(x', y'), \quad y(0) = 0, \quad y'(0) = c_2 .$$

Introducing, as before, the function $f(c_1, c_2)$ equal to the maximum altitude, we see that

$$(11.5.2) \qquad f(c_1, c_2) = (c_1^2 + c_2^2)^{1/2} \Delta$$

$$+ f\left[c_1 + g(c_1, c_2)\Delta, c_2 + h(c_1, c_2)\Delta\right] + 0(\Delta) .$$

Hence,

$$(11.5.3) \qquad (c_1^2 + c_2^2)^{1/2} + g(c_1, c_2) \frac{\partial f}{\partial c_1} + h(c_1, c_2) \frac{\partial f}{\partial c_2} = 0 .$$

Once again, let us assume that $c_2 = 0$ implies no vertical motion. Then $f(c_1, 0) = 0$ for $c_1 \geq 0$. It follows that we can again compute the solution by means of a sequence of functions of one variable.

11.6 Maximum Range

To tackle the problem of maximum range directly requires the introduction of another state variable, the initial altitude. It can also be broken up into two problems, corresponding to the ascent to maximum altitude, and the descent.

11.7 Maximum Penetration

Let us point out that the same methods can be used to obtain the maximum penetration of a high energy particle in a shield as a function of the initial velocity and angle.

Bibliography and Comments

Section 11.1. We are following the paper, Bellman, R., "Functional Equations and Maximum Range", Quarterly of Applied Mathematics, Vol. XVII, No. 3, October, 1959, pp. 316-318.

Chapter XII

SEMIGROUPS IN SPACE

12.1 Introduction

Invariant imbedding is a mathematical theory devoted to the exploitation of structural features of processes. In consequence, it is a loose confederation of ideas and techniques, of methods and methodology which can be employed in the analytic and computational study of large classes of mathematical and scientific questions. In what follows we will discuss some basic analytic aspects of the theory and provide references to more detailed analysis and numerical results.

We can begin with the observation that problem solving is a principal occupation of the intellectual. A powerful procedure widely employed in this pursuit is "imbedding". By this term we mean that the procedure whereby the resolution of a specific question is accomplished by consideration of a family of related questions. Rather remarkably, it turns out that it is often far easier to treat a set of problems in unison rather than a single problem in isolation. This is the essence of the comparative method familiar to so many disciplines: comparative linguistics, comparative anatomy, comparative religion, comparative anthropology, to name a few. Perhaps needless to say, it is not always an easy matter to discern the connecting links and thus an appropriate family. Banach is reputed to have said that brilliance consists of spotting analogies, and genius of seeing analogies between analogies.

The methods that are discussed below can be applied to many parts of mathematical physics and control theory, to mathematical economics, scheduling theory and operations research. They constitute an important part of the modern mathematical approach to the study of systems.

12.2 Imbedding in Time

Let us begin with a classical example of the method. Suppose that we are given the current state of a system and asked to predict the state at some subsequent time T. One way to go about this is to consider the general problem of predicting the state of the system at any subsequent time $t > 0$, where 0 denotes the present time.

We begin then by introducing a function $x(t)$, the state of the system at time t. For our present purposes we assume that this is a finite dimensional vector of dimension N. The next step is to obtain relations between functional values of x for different values of t. In many cases we can obtain an equation of the form

$$(12.2.1) \qquad x(t+\Delta) = x(t) + g(x(t))\Delta + \dots$$

for small positive Δ. In the limit as $\Delta \to 0$ this yields the differential equation

$$(12.2.2) \qquad x'(t) = g(x(t)) \quad ,$$

with an initial condition $x(0) = c$.

The prediction problem has been transformed into the task of solving a functional equation.

12.3 Advantages and Disadvantages

This is a very powerful and flexible method which has had widespread success in science. It can be made the basis of numerous computational algorithms, algorithms which can be quickly and accurately carried out with the aid of digital computers. There are, however, some drawbacks as, of course, there must be to every method. In a number of cases too much data is calculated at too high a cost in both time and accuracy.

One way to circumvent these difficulties is to use some semigroup properties of the process. From the physical point of view this means taking advantage of the law of causality; from the mathematical point of view it means exploiting existence and uniqueness of solution. The impetus to this approach is due to Hadamard.

If (12.2.2) is a linear equation,

$$x'(t) = Ax , \qquad x(0) = c \quad , \tag{12.3.1}$$

the semigroup property is made apparent using the exponential form of the solution,

$$x = e^{At}c \quad . \tag{12.3.2}$$

We see that

$$e^{A(s+t)} = e^{As}(e^{At}) \quad . \tag{12.3.3}$$

This allows us to use doubling techniques,

$$e^{2At} = (e^{At})^2; \quad e^{2^N t} = (e^t)^{2^N} \quad . \tag{12.3.4}$$

Thus N successive squarings will yield $e^{2^N t}$ starting with e^t. This is a considerable acceleration time. Nonetheless, there remain many interesting questions connected with the calculation of e^{NAt} given e^{At}.

12.4 Iteration

If (12.2.2) is nonlinear, we must use the more general approach of iteration to illustrate the underlying semigroup properties. Write

$$(12.4.1) \qquad x(t) \equiv x(t, c) = f(c, t) \quad .$$

Then uniqueness of solution (assuming that $g(x)$ is well-behaved, e.g. analytic) yields that basic semigroup relation

$$(12.4.2) \qquad f(c, s+t) = f(f(c, s), t) \ , \quad s, t > 0 \quad ,$$

with $f(c, 0) = 0$. In some cases this approach, together with the concept of relative invariants can be used to accelerate the calculation of $f(c, T)$.

The semigroup determined by the linear equation of (12.3.1) generalizes in several ways. One generalization is afforded by a nonlinear differential equation as the basic equation; one is provided by two-point boundary value problems of the type discussed subsequently in place of an initial value problem; one is provided by the theory of multistage decision processes, which is to say dynamic programming, in place of a descriptive process. There the equation is quasilinear, namely

$$(12.4.3) \qquad x'(t) = \max_q \left[A(q)x(t) + b(q)\right] \quad ;$$

In this fashion the calculus of variations is imbedded in semigroup theory.

12.5 Imbedding in Space

Let us consider a further, and equally important, example of the imbedding method. Consider a steady-state transport process in a one-dimensional rod with an incident flux c at one end point T. (see Fig. 12.1)

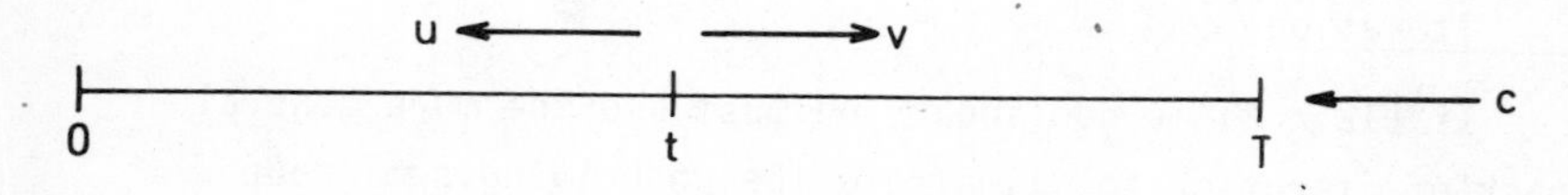

Fig. 12.1

We are asked to determine the reflected and transmitted fluxes, under various assumptions concerning the interaction of the flux with the medium and with itself. To answer this question using the technique of imbedding, we enlarge the investigation by asking for the values of the left-hand and right-hand fluxes, $u(t)$ and $v(t)$, at any interior point t. The quantity $v(T)$ is the desired reflected flux; $u(0)$ is the required transmitted flux.

Examination of the relations between $u(t)$, $u(t \pm \Delta)$, $v(t)$ and $v(t \pm \Delta)$ (local conservation relations) yields a pair of differential equations

$$
\begin{aligned}
(12.5.1) \qquad u'(t) &= g(u(t), v(t)) ; \qquad u(T) = c \quad , \\
v'(t) &= h(u(t), v(t)) ; \qquad v(0) = 0 \quad .
\end{aligned}
$$

Observe that this is a two-point boundary value problem. We have insufficient information at $t = 0$ and $t = T$ to resolve the equation as an initial problem.

12.6 Advantages and Disadvantages

If we can solve (12.5.1), we will have obtained a solution of the original problem as well as a good deal of additional information of interest. However, a serious drawback to this approach lies in the fact that this equation cannot be used to provide a guaranteed algorithm for a digital computer the way an initial value equation can. Two-point boundary value problems are notoriously difficult, both analytically and computationally.

This obstacle, as well as the time barrier, may be turned by the use of the hybrid computer, analog plus digital. Since these questions, however, have not been investigated to any extent, we shall say no more at this point.

12.7 An Imbedding in Structure

Let us now imbed the original questions, the determination of the reflected and transmitted fluxes in a different family of problems. Let us seek to obtain these desired fluxes as functions of the initial intensity and the thickness of the rod.

To this end we write the reflected and transmitted fluxes.

$$(12.7.1) \qquad v(t) = r(c, T) \quad ,$$

$$u(0) = t(c, T) \quad ,$$

as functions of these parameters.

To obtain equations for these functions we use some semi-group ideas. It turns out that for this purpose it is convenient to introduce an additional variable of physical significance. Suppose that a flux d is incident from the left at 0, as indicated below (Fig. 12.2).

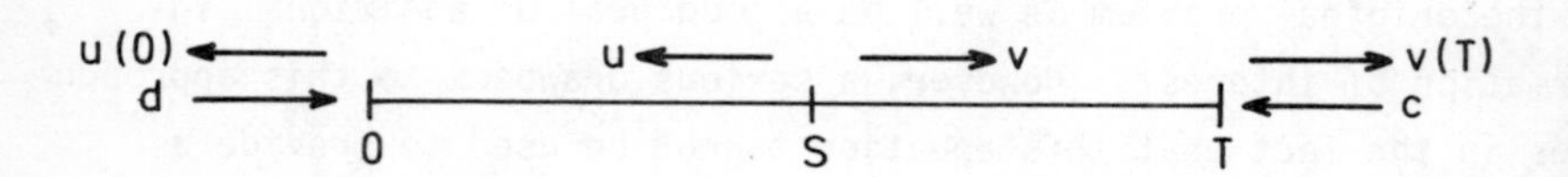

Fig. 12.2

Let then

$$(12.7.2) \qquad v(T) = r(c, d, T) \quad ,$$

$$u(0) = t(c, d, T) \quad ,$$

denote the respective "reflected" and "transmitted" fluxes, which is to say the fluxes emergent from the right and left, as indicated.

Considering the interval $[0, S]$, we may regard $v(S)$ as the reflected flux due to fluxes d and $u(S)$. Thus

$$v(S) = r(u(S), d, S) \quad . \tag{12.7.3}$$

Similarly, the interval $[S, T]$ yields the relation

$$u(S) = t(c, v(s), T - S) \quad . \tag{12.7.4}$$

We also have, using the entire interval,

$$\begin{aligned} r(c, d, T) &= r(c, v(S), T - S) \quad , \\ t(c, d, T) &= t(u(S), d, T) \quad . \end{aligned} \tag{12.7.5}$$

Elimination of $u(S)$ and $v(S)$ yields the basic semigroup relations for the functions $r(c, d, T)$ and $t(c, d, T)$.

12.8 Associated Partial Differential Equation

The limiting form of the foregoing relations as $S \to T$ are partial differential equations for the functions $r(c, d, T)$ and $t(c, d, T)$. These are of initial value type since the reflection and transmission functions are simply specified for $T = 0$, namely

$$\begin{aligned} r(c, d, 0) &= d \quad , \\ t(c, d, 0) &= c \quad . \end{aligned} \tag{12.8.1}$$

This is quite interesting since it shows that the two-point boundary value problem for an ordinary differential equation can be solved in terms of an initial value problem for a partial differential equation, and conversely. This of course is connected with characteristic theory. This flexibility is important for both analytic and computational reasons.

If (12.8.1) is linear, then

$$r(c, d, T) = R_1(T)c + T_1(t)d \quad , \tag{12.8.2}$$

$$t(c, d, T) = T_1(T)c + R_1(T)d \quad ,$$

where R_1 and T_1 are reflection and transmission matrices. Here we are assuming both homogeneity and isotropy. This need not be the case, requiring the introduction of additional matrices. The functional equations of (12.7.3), yield addition formulas for these matrices, and the partial differential equations then yield the familiar Riccati equations for the reflection and transmission matrices.

12.9 Imbedding in Structure

Let us give another example of imbedding in structure associated with the potential equation. Let R be a region with boundary B, (Fig. 12.3).

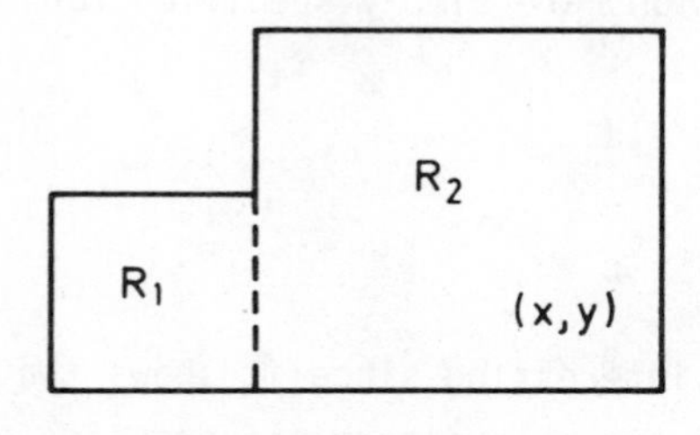

Fig. 12.3

Consider the equation

$$(12.9.1)\qquad u_{xx} + u_{yy} = 0 , \quad (x, y)\varepsilon R ,$$

$$u = g(x, y) , \quad (x, y)\varepsilon B .$$

The usual determination of u at a particular point (x_1, y_1) depends upon a simultaneous determination of u for all (x, y) in R. We can approximate to this problem by using a suitable subset of points and thus reduce the problem to a finite one.

Can we use a "building block" method to resolve the problem for $R = R_1 + R_2$ based upon the subregions R_1 and R_2 of simpler nature? It turns out that this can be done using ideas of invariant imbedding or of dynamic programming.

12.10 Applications in Mathematical Physics

We have mentioned above an application of invariant imbedding to transport theory, the origin of all of the subsequent investigations. It was the pioneering work of Ambarzumian in the theory of radiative transfer and then the deeper and more extensive research of Chandrasekhar, his "principles of invariance", which made it clear that a vast domain of mathematical analysis awaited for exploration.

It turns out that there are many processes in mathematical physics whose structures are ideally suited to invariant imbedding. Let us mention random walk, wave propagation and analytical mechanics.

12.11 Application to Combinatories

In the foregoing, a very simple partitioning, or stratification, technique was used, being applied to both intervals and regions. We can also employ a partitioning of sets. This enables us to treat a number of combinatorial problems by means of the functional equation of dynamic programming.

12.12 Discussion

The foregoing studies lead to the posing of a large number of important, interesting and difficult questions. Some of these are:

1. In how many ways may we meaningfully imbed a particular process in a family of processes and how do we determine these imbeddings?
2. How do we calculate desired data with as little extraneous calculation as possible?
3. How do we minimize the time required to calculate desired data?
4. How does one obtain the desired data using algorithms of specified type? This is a generalized Mascheroni construction.
5. What is the minimum data set required to obtain a specified value?
6. What are the simplest possible analytic structures which will yield the desired reşults?

Bibliography and Comments

Section 12.1. We are following the paper, Bellman, R., "Invariant Imbedding, Semigroups in Time, Space and Structure". Conference on Applications of Analytic Numerical Analysis, Vol. 228, Springer-Verlag, 1971, pp. 9-18.

For an introduction to invariant imbedding, see the book, Bellman, R., and Wing, G. M., An Introduction to Invariant Imbedding, John Wiley & Sons, Inc., New York, 1974.

Section 12.11. See the book, Bellman, R., Cooke, K. L., and Lockett, J., Algorithms, Graphs and Computers, Academic Press, New York, 1970.

Chapter XIII
VARIATIONAL PROBLEMS AND FUNCTIONAL EQUATIONS

13.1 Introduction

The purpose of this chapter is to indicate some interesting functional equations arising in the minimization of quadratic functionals. The method is quite general, as we shall discuss in Sec. 13.3, but to illustrate it we shall treat the following problem: Minimize the functional

$$J(u) = \int_r^s (u'^2 + g(t)u^2)dt \quad , \tag{13.1.1}$$

where u is subject to the conditions

$$u(r) = a \;, \qquad u(s) = b \quad . \tag{13.1.2}$$

Let us write

$$f(r, s, a, b) = \min_u J(u) \quad . \tag{13.1.3}$$

We know that f is a quadratic function of a and b. We have

$$f(r, s, a, b) = a^2 f_1(r, s) + abf_2(r, s) + b^2 f_3(r, s) \quad . \tag{13.1.4}$$

Let us assume that $g(t)$ is positive so that the minimum exists for all r and s. We know, in this case, that f is positive definite.

13.2 Derivation of the Functional Equations

Let p be a point between r and s. Set

$$u(p) = b \quad . \tag{13.2.1}$$

$$f(r, b, a, c) = \min_{c} \left[f(r, p, a, c) + f(p, s, c, b) \right] \quad . \tag{13.2.2}$$

The minimization is easily obtained in view of the quadratic character of f. Equating the coefficients, we obtain the desired functional equations.

13.3 Some Generalizations

The result may easily be generalized in various ways. Let us point out some of them.

To begin with we can consider functionals involving higher derivatives, and we can consider the vector-matrix form. We may also use several points between r and s. We can consider non-quadratic integrands but it is now not easy to determine the form of f or to perform the minimization.

We may also employ sums rather than integrals. The derivative is now replaced by the difference. We may also consider other quadratic functionals such as those connected with linear integral equations and with non-local phenomena.

Finally, we may consider multidimensional cases. Here, the analytic details are more complex, although the basic idea is the same.

Bibliography and Comments

Section 13.1. The minimization of quadratic functions is discussed in detail in Bellman, R., *Introduction to the Mathematical Theory of Control Processes*, Vol. I, Academic Press, Inc., New York, 1968.

Chapter XIV

ALLOCATION PROCESSES, LAGRANGE MULTIPLIERS AND THE MAXIMUM TRANSFORM

14.1 Introduction

In this chapter we consider allocation processes. These processes arise whenever we face the task of apportioning resources among activities.

The first approach we give depends on the Lagrange multiplier We show that the Lagrange multiplier furnishes a very important reduction of dimensionality. It replaces a resource constraint by a price.

Then we consider the maximum transform. We give the necessary foundations for the analysis.

In the final section, we consider some connections between the maximum transform and semigroups.

14.2 Lagrange Multipliers

The purpose of this section is to indicate how a suitable combination of the classical method of the Lagrange multiplier and the functional-equation method of the theory of dynamic programming can be used to solve numerically, and treat analytically, a variety of variational problems that cannot readily be treated by either method alone.

Consider the problem of maximizing the function

$$F(x_1, x_2, \ldots, x_N) = \sum_{i=1}^{M} g_i(x_i) \quad , \tag{14.2.1}$$

subject to the constraints

$$\sum_{j=1}^{N} a_{ij}(x_j) \leq c_j \; , \qquad i = 1,2,\ldots, M \quad , \tag{14.2.2a}$$

$$x_i \geq 0 \tag{14.2.2b}$$

where the functions $a_{ij}(x)$, $g_i(x)$ are taken to be continuous for $x \geq 0$, and monotone increasing. For $c_i \geq 0$, define the sequence of functions

$$f_N(c_1, c_2, \ldots, c_M) = \underset{[x]}{\text{Max}}\; F(x_1, x_2, \ldots, x_N) \tag{14.2.3}$$

for $N \geq 1$.

Then $f_i(c_1, c_2, \ldots, c_M)$ is determined immediately, and employing the principle of optimality, we obtain the recurrence relation

$$f_{k+1}(c_1, c_2, \ldots, c_M) = \underset{0 \leq a_{i,k+1}(x) \leq c_i}{\text{Max}} [g_{k+1}(x) + f_k(c_1 - a_{1,k+1}(x), \ldots, c_M - a_{M,k+1}(x))] \quad , \tag{14.2.4}$$

for $k = 1,2,\ldots, N-1$.

Due to the limited storage of present-day digital computers, this method flounders on the reef of dimensionality when $M \geq 4$. If we wish to treat applied problems of greater and greater realism, we must develop methods capable of handling problems involving higher dimensions.

In these sections we shall present one method of overcoming these dimensionality difficulties.

The method of the Lagrange multiplier in classical variational theory consists of forming the function

$$(14.2.5) \qquad \phi(x_1, x_2, \ldots, x_N) = \sum_{i=1}^{N} g_i(x_i) - \sum_{i=1}^{M} \lambda_i \left(\sum_{j=1}^{N} a_{ij}(x_j) \right)$$

where the λ_i are parameters subsequently found by means of relations (14.2.2), and then utilizing a direct variational on this new function.

We shall employ an approach intermediate between this method and the method sketched above.

Consider the function

$$(14.2.6) \qquad F(x_i, x_2, \ldots, x_N; \lambda_1, \lambda_2, \ldots, \lambda_K) = \sum_{i=1}^{N} g_i(x_i) - \sum_{i=1}^{K} \lambda_i \left(\sum_{j=1}^{N} a_{ij}(x_j) \right) ,$$

where $1 \leq K \leq M - 1$. We wish to maximize this function over the region defined by the constraints

$$(14.2.7a) \qquad \sum_{j=1}^{N} a_{ij}(x_j) \leq c_i , \qquad i = K + 1, \ldots, M ,$$

$$(14.2.7b) \qquad x_i \geq 0 .$$

For fixed values of the λ_i, we have a problem of precisely the type discussed above, with the advantage that we now require functions of dimension $M - K$ for a computational solution.

Once the sequence $\{\phi_N(c_{K,1}, \ldots, c_N; \lambda_1, \lambda_2, \ldots, \lambda_K)\}$, $N = 1,2,\ldots,$

$$(14.2.8)\qquad \phi_N(c_{K,1},\ldots, c_N; \lambda_1, \lambda_2,\ldots, \lambda_K) = \underset{[x]}{\text{Max}}\ F(x_1, x_2,\ldots, x_N; \lambda_1, \lambda_2,\ldots, \lambda_K)$$

has been computed, we vary the parameters λ_i to determine the range of the parameters $c_1, c_2,\ldots, c_K$.

We have thus partitioned the computation of the original sequence of functions of M variables into the computation of a sequence of functions of K variables followed by the computation of a sequence of functions of $M-K$ variables. The choice of K will depend upon the process.

14.3 Successive Approximations

Since the Lagrange multipliers are intimately connected with "marginal returns", or "prices", in a number of applied problems, we begin with a certain hold on the process as far as approximate solutions are concerned. Iterative techniques based upon this observation, and the connection with optimal search procedures can be obtained.

14.4 Application to the Calculus of Variations

In the theory of control processes, one encounters a problem such as that of minimizing a nonanalytic functional such as

$$(14.4.1)\qquad J_1(u) = \int_0^T |1 - u|\ dt$$

over all functions $v(t)$ satisfying the constraints

$$(14.4.2a)\qquad -k_1 \le v(t) \le -k_2\ ,\qquad 0 \le t \le -T\quad ,$$

$$(14.2.2b)\qquad \int_0^T |v(t)|\ dt \le c_1\quad ,$$

where u and v are connected by a relation

$$\frac{du}{dt} = g(u, t, v), \quad u(0) = a \quad . \tag{14.4.3}$$

Replacing $J_1(u)$ by

$$J_2(u) = \int_0^T |1 - u| \, dt + \lambda \int_0^T |v(t)| \, dt \quad , \tag{14.4.4}$$

$\lambda \geq 0$, we can employ functions of two variables in determining the analytic or numerical solution, rather than functions of three variables. This reduction by one in dimensionality results simultaneously in a tremendous saving in computing time and in a great increase in accuracy of the numerical results.

14.5 Maximum Transform

A functional operation that occurs in a variety of mathematical programming (optimization) problems of economics and operation research is

$$h(x) = \max_{u+v=x} [f(u) + g(v)] \quad ; \tag{14.5.1}$$

here, f and g are real-valued, continuous functions of $x = (x_1, x_2, \ldots, x_n)$ on the domain $x_i \geq 0$, $i = 1,2,\ldots, n$, and u, v, x lie in the domain. We term h the maximum convolution of f, g and symbolize (14.5.1) by

$$h = f \oplus g \quad . \tag{14.5.2}$$

It would be of considerable value for the solution of problems involving this binary operation to determine and study functional transformations T which convert $\oplus$ to ordinary addition, that is, which have the "disassociative" property

$$T(f \oplus g) = Tf + Tg \quad . \tag{14.5.3}$$

It is this question that the present sections are devoted to.

To relate this question to more familiar terrain, it is convenient to formulate (14.5.2) and (14.5.3) in an equivalent multiplicative form (see Sec. 14.11). Let the (multiplicative) maximum convolution of nonnegative function F, G be defined by

$$(14.5.4) \qquad H(x) = (F \quad G)(x) = \max_{u+v=x} [F(u) \times G(v)] \qquad ;$$

then in place of (14.5.3), we consider

$$(14.5.5) \qquad T(F \otimes G) = (TF) \times (TG) \qquad .$$

Now an analog of (14.5.4) is ordinary convolution

$$(F * G)(x) = \int_0^x F(y)\, G(x - y)\, dx \qquad ,$$

and property (14.5.5) in this case is a key feature of the Laplace operator

$$\int_0^\infty e^{-tx}\, F(x)\, dx \qquad .$$

the transformations T of (t), or (14.5.3), occupy a role in the study of the maximum convolution comparable in many ways to that of the Laplace transform in ordinary convolution. The analogy is even more explicit, as we shall see in Sec. 14.11, the counterpart of the Laplace transform given by

$$(14.5.6) \qquad \Phi(\xi) = (TF)(\xi) = \max_{x \geq 0} [e^{-\xi x} F(x)]$$

does, in fact, satisfy (14.5.5). Further, the transformation

$$F(x) = (T^{-1}\Phi)(x) = \min_{\xi} [e^{r\xi} \Phi(\xi)]$$

serves as an inverse to T. For applications, it seems preferable to formulate the theory in the additive form of (14.5.2) and (14.5.3), and that course will be followed here.

The chapter is summarized as follows. In Sec. 14.6 we

organize several definitions and facts concerning convexity in a manner which is suited to the present purposes; in particular, we introduce the concave increasing cap, or envelope, $\bar{f}$ or a function f. In Sec. 14.7, we define the maximum transform $\phi = Mf$ by the relation

$$(14.5.7) \qquad -\phi(\xi) = \sup_{x>0} \{ - (\xi, x) + f(x) \} \quad ;$$

here, $x \geq 0$ denotes $x_i \geq 0$ for all i, and $(\xi, x) = \sum_{i=1}^{n} \xi_i x_i$. This is a counterpart of (14.5.6) for the additive case. It is shown that M has property (14.5.3), $Mf = M\bar{f}$, and $M\phi = \bar{f}$. In Secs. 14.8 and 14.9, two applications of the maximum transform are given; one to the problem of optimal distribution of effort, and another to a multistage allocation process.

In Sec. 14.10, the general transformation T satisfying (14.5.3) is considered; it is shown that $Tf = T\bar{f}$ and T factors into λM, where the transformation λ is additive in that $\lambda(\phi_1 + \phi_2) = \lambda\phi_1 + \lambda\phi_2$. Thus, any T must identify f and g if $\bar{f} = \bar{g}$, and M is a "best" transformation T in the sense that $Mf \neq Mg$ whenever $\bar{f} \neq \bar{g}$. In Sec. 14.11 the modifications required in M are given for the multiplicative transform and the minimum convolution. In Sec. 14.12, a short list of maximum transforms is presented.

14.6 Definitions

We use x, ξ, etc., to denote n-tuples of real numbers, representing points in Euclidean n-space; z denotes a real number and (x, z) a point in $(n + 1)$-space. The inequality $x \geq 0$ means that each component of $x = (x_1, x_2, \ldots, x_n)$ satisfies $x_i \geq 0$ while $x = 0$ denotes the origin $(0,0,\ldots,0)$. We consider real-valued functions f which are defined (finite) for every $x \geq 0$. With f we associate the set in $(n+1)$-space $|f| = \{(x, z) \mid z \leq f(x)\}$. The function f is concave in case $f(px + qy) \geq pf(x) + qf(y)$ for any $x \geq 0$ and $y \geq 0$ and any real numbers $p \geq 0$, $q \geq 0$ with $p + q = 1$

(the inequality is reversed for convex functions); this holds just in case $|f|$ is a convex point set in $(n+1)$-space.

We assume a knowledge of various fundamental properties of convex sets in Euclidean space. Pertinent definitions are given here, for the reader's convenience. For $(u, w) \neq 0$, the equation

$$(14.6.1) \qquad wZ + (u, X) = b$$

will denote the hyperplane in $(n+1)$-space, where (X, Z) are running coordinates having the direction (u, w) as its "positive" normal direction. The half-space bounded by (14.6.1) will be selected as the set of points (x, z) satisfying

$$(14.6.2) \qquad wz + (u, x) \leq b$$

and we refer to (u, w) as the direction of either (14.6.1) or (14.6.2). When $w = 0$, the hyperplane is "vertical"; when $w > 0$ the normal is "upward" and in this case (14.6.2) is equivalent to

$$(14.6.3) \qquad z < (\xi, x) + b$$

with normal direction $(-\xi, 1)$; if $\xi \geq 0$, we say that the hyperplane (or its half-space) "slants upward".

A point set S in $(n+1)$-space has (14.6.1), or (14.6.2), as a barrier in case S lies in the half-space (14.6.2), that is

$$(14.6.4) \qquad wz + (u, x) \leq b \,, \qquad (x, z) \in S \quad .$$

A barrier of S is a support in case b is the least value for which (14.6.4) holds, that is,

$$(14.6.5) \qquad \sup_{(x,z)\in S} \{wz + (u, x)\} = b \quad .$$

The convex hull of S is the smallest convex set containing S.

The function f is said to have (14.6.1) as a barrier (support) in case the set $|f|$ has (S) as a barrier (support).

From the definition $|f|$, it follows that every barrier of f has $w \geq 0$; in this case (14.6.4) and (14.6.5) are equivalent, respectively, to

$$(14.6.6) \qquad wf(x) + (u, x) \leq b , \qquad x \geq 0 \quad ,$$

$$(14.6.7) \qquad \sup_{x \geq 0} \{wf(x) + (u, x)\} = b \quad .$$

That f is bounded in the direction (u, w) means that there is a barrier of f (and hence a support) having this direction; this means that the projections of $(x, f(x)), x \geq 0$, on the direction (u, w) are bounded in this direction.

An admissible function is a function which is continuous on $x \geq 0$ and bounded in any direction $(-\xi, 1)$ with $\xi_i > 0$, $i = 1,2,\ldots, n$. Thus, a continuous function on $x \geq 0$ is admissible in case it is dominated by any positive multiple kr of the length r of x, as $r \to \infty$. If f, g are admissible, so is $f \oplus g$; if, further, f, g are concave, so is $f \oplus g$. In this section, the convolution and transformations (14.6.2) and (14.6.3) will be considered for such functions; functions on $x \geq 0$ are understood to be admissible unless stated otherwise.

The function f is increasing in case $f(x) \geq f(y)$ whenever $x \geq y$. In applications, $f(x)$ ordinarily measures the return resulting from an input of x_i units of each of n resources i; such a function would be expected to be increasing. For greater generality however, we do not directly restrict f in this way, but associate with f the admissible increasing function

$$(14.6.8) \qquad f^+(x) = \max_{0 \leq y \leq x} f(y) \quad .$$

Note that for every $x \geq 0$, there exists a $y \geq 0$ such that

$$(14.6.9) \qquad f^+(x) = f(y) , \qquad y \leq x \quad .$$

We now define the concave increasing envelope $\bar{f}$ of f as the "cap" of the convex hull H of the point set $[f^+]$. That is,

$$\bar{f}(x) = \sup_{(x,z)\in H} z \; ; \tag{14.6.10}$$

the points $(x, \bar{f}(x))$ belong to the closure of H. The function $\bar{f}$ is concave, increasing and admissible. Finally, for $N = 1,2,3,\ldots,$ let f_N^+ denote f^+ restricted to $0 \le x_i \le N$, $i = 1,2,\ldots, n$,

$$f_N^+(x) = f^1(x) \; , \qquad 0 \le x_i \le N \quad ,$$

and let H_N = convex hull of $[f_N^+]$. This completes the necessary definitions.

<u>LEMMA 1</u>. Every support of $[f^+]$ is either vertical or slants upward. Also, a half-space which slants upward is a support of $[f^+]$ if and only if it is a support of $[f]$. In particular, $[f^+]$ and $[f]$ have the same support for any direction $(-\xi, 1)$ with $\xi_i > 0$, $i = 1,2,\ldots, n$, and such a support makes contact with $[f]$ at finite point $(x, f(x))$.

<u>PROOF</u>. If (u, w) is the direction of a support of $[f^+]$, then $w \ge 0$, since $[f^+]$ contains $(x, z) = (x, -k)$ for arbitrarily large positive k. Suppose $u_j > 0$ for some j. With each (x, z) the set $[f^+]$ also contains $(x^*, z) = (x_i,\ldots, x_j + k,\ldots, x_n, z)$ for arbitrary $k > 0$. Then $wz + (u, x^*)$ is unbounded from above as $k \to \infty$, contrary to (4). Hence $w \ge 0, u \le 0$; this proves the first statement of the lemma.

Consider a half-space which slants upward and is a barrier of $[f]$; let $(-\xi, 1)$ be its direction, with $\xi \ge 0$. Then for some b

$$f(x) \le (\xi, x) + b \; , \qquad x \ge 0 \quad .$$

Consider $f^+(x)$; by (14.6.9)

$$f^+(x) = f(y) \le (\xi, y) + b \le (\xi, x) + b \; , \qquad x \ge 0 \quad .$$

Hence the half-space is also a barrier of $[f^+]$. Conversely, every barrier of $[f^+]$ is a barrier of $[f]$, since $f(x) \le f^+(x)$. This proves the second statement of the lemma.

To prove the third statement note that $(x^*, f(x^*))$ is a point of contact in case the maximum of $f(x) - (\xi, x)$ is realized at x^*; but there is such an x^* since $f(x) - (\xi/2, x)$ is bounded above and, accordingly, $f(x) - (\xi, x) \to -\infty$ if any $x_i \to \infty$.

LEMMA 2. The closure of H is the intersection of all supports of $[f^+]$ (or of $[f]$) which slant upward. It follows that

$$\bar{f}(x) = \inf_{\xi} \{(\xi, x) + b(\xi)\} \quad , \qquad x \ge 0 \quad ,$$

where $\xi \ge 0$ ranges over those directions $(-\xi, 1)$ in which f is bounded, and $b(\xi)$ is the z-intercept of the corresponding support hyperplane.

PROOF. The set H is the convex hull of $[f^+]$; it is a standard result in convex sets that the closure $\bar{H}$ of H is the intersection of the supports of $[f^+]$. By Lemma 1, it is only necessary to show that the vertical supports may be dispensed with; this may be done by proving that any point outside of $\bar{H}$ can be separated from $\bar{H}$ by a nonvertical barrier of $\bar{H}$. This completes the proof.

14.7 Analytic Properties of the Maximum Transform

We now turn to a consideration of the maximum transform

$$\phi = Mf$$

defined by (14.5.7) for admissible functions f. The domain of definition D of ϕ is taken as all $\xi \ge 0$ such that f is bounded in the direction $(-\xi, 1)$; in particular, by Lemma 1, D contains those ξ with all components $\xi_i > 0$, and for such a ξ

$$(14.7.1)\qquad -\phi(\xi) = \max_{x \geq 0} \{-(\xi, x) + f(x)\}, \quad \xi > 0 \quad .$$

<u>THEOREM 1</u>. The maximum transformation M satisfies $M(f \oplus g) = Mf + Mg$ for admissible functions f and g, the domain of definition of $M(f \oplus g)$ being the domain common to Mf and Mg. The value $-\phi(\xi)$ is the z-intercept of the support hyperplane of f having the direction $(-\xi, 1)$. If ξ belongs to the domain D of ϕ then so does every $\eta \geq \xi$ and $\phi(\eta) \geq \phi(\xi)$; that is ϕ is an increasing function; also, ϕ is bounded on D.

<u>PROOF</u>. If f and g are both bounded in the direction $(-\xi, 1)$ then so is $f \oplus g$, and conversely; hence the domain of $M(f \oplus g)$ is the intersection of those of Mf and Mg. Furthermore,

$$\begin{aligned}
-[M(f \oplus g)](\xi) &= \sup_{x \geq 0} \{-(\xi, x) + \sup_{0 \leq y \leq x} [f(y) + g(x - y)]\} \\
&= \sup_{x \geq 0} \sup_{0 \leq y \leq x} \{-(\xi, x) + f(y) + g(x - y)\} \\
&= \sup_{y \geq 0} \sup_{x \geq y} \{-(\xi, x) + f(y) + g(x - y)\} \\
&= \sup_{y \geq 0} \sup_{x \geq y} \{[-(\xi, x) + f(y)] + [-(\xi, x - y) + g(x - y)]\} \\
&= \sup_{y \geq 0} \sup_{u \geq 0} \{[-(\xi, y) + f(y)] + [-(\xi, u) + g(u)]\} \\
&= \sup_{y \geq 0} \{-(\xi, y) + f(y)\} + \sup_{u \geq 0} \{-(\xi, u) + g(u)\} \\
&= -(Mf)(\xi) - (Mg)(\xi) \quad .
\end{aligned}$$

This proves the first statement of the theorem. To verify the second statement notice that $-\phi(\xi)$ is identical with b of (14.6.5), where $(u, w) = (-\xi, 1)$; from (S), b is the z-intercept of the supporting hyperplane. The first part of the last statement is immediate from (14.5.7), the second part from the fact that the z-intercept of the support hyperplane is always $\geq f(0)$.

<u>THEOREM 2</u>. Let f be an admissible function; let f^+ and $\bar{f}$ be its associated increasing and concave increasing function respectively. Then

$$\phi = Mf = Mf^+ = M\bar{f} \quad ,$$

with these transforms having a common domain of definition D. The set D is convex, contains the open set $\{\xi|\xi_i > 0\}$, and contains every "corner" $\eta \geq \xi$ with ξ in D. The transform ϕ is a concave increasing function on D.

<u>PROOF</u>. That the three transforms have the common domain D in $\xi \geq 0$ and are equal follow from the fact that they have the same supports with directions $(-\xi, 1)$ by Lemmas 1 and 2. The convexity of D and the concavity of ϕ are direct consequences of (14.5.7) and the definitions of these properties. The remaining points of the theorem repeat preceding comments.

The operator M may be applied to $\phi(\xi)$ as follows,

$$(14.7.2) \qquad (M\phi)(x) = -\sup_{\xi \in D} \{-(x, \xi) + \phi(\xi)\}$$

From Lemma 2, it follows that the domain of $M\phi$ is $\{x|x \geq 0\}$ and that

$$(14.7.3) \qquad M\phi = \bar{f} \quad .$$

<u>THEOREM 3</u>. For any admissible function f

$$(14.7.4) \qquad M(Mf) = \bar{f} \quad .$$

Consequently,

$$\bar{f} = \bar{g} \text{ if and only if } Mf = Mg.$$

<u>PROOF</u>. Equation (14.7.4) is a restatement of (14.7.3). If

$= \bar{g}$, then $M\bar{f} = M\bar{g}$, or $Mf = Mg$; conversely, if $Mf = Mg$, then $(Mf) = M(Mg)$, or $\bar{f} = \bar{g}$.

Equation (14.7.3) is the inversion formula for equation $f = \phi$, yielding the solution for f in terms of the transform ϕ. n applications, use of the operator M leads to solutions "modulo" he concave increasing envelope of the desired function.

Some useful features of M are the following: If

$$\phi = Mf \; , \qquad \psi = Mg$$

hen for scalars $a > 0$, $b > 0$ and c,

$$\text{(14.7.5)} \qquad \phi(\xi) - c = M\{f(x) + c\}\,(\xi) \quad ,$$

$$b\phi\left(\frac{\xi}{ab}\right) = M\{bf(ax)\}\,(\xi) \quad .$$

his is a consequence of the definition (14.7.5), $M[\;](\xi)$ denoting he value at ξ of the transform of the function appearing in the rackets. The functions ϕ, f (and variables ξ, x) can be interchanged in (14.7.5) when $f = M\phi$, that is, when f is concave increasing. Notice that $\overline{f \oplus g} = \bar{f} \oplus \bar{g}$; this follows from $\phi = Mf = M\bar{f}$, $\psi = Mg = M\bar{g}$ and $\overline{f \oplus g} = M[M(f \oplus g)] = M(\phi + \psi) = M[M(\bar{f} \oplus \bar{g})] = \bar{f} \oplus \bar{g}$, this last function being concave increasing with $\bar{f}$ and $\bar{g}$.

Points $(x, f(x))$ of the graph of f and points $(\xi, \phi(\xi))$ of its transform satisfy the relation $(\xi, x) \geq f(x) + \phi(\xi)$. Let f be concave increasing; the following is a consequence of the preceding discussion. If $(-\bar{\xi}, 1)$ is a support direction of f at $(\bar{x}, f(\bar{x}))$, then $(-\bar{x}, 1)$ will be a support direction of ϕ at $(\bar{\xi}, \phi(\bar{\xi}))$, and $x = \bar{x}$, $\xi = \bar{\xi}$ satisfy the equation

$$\text{(14.7.6)} \qquad (\xi, x) = f(x) + \phi(\xi) \quad .$$

When the function f is sufficiently smooth, then (14.7.6) together with the transformation of coordinates

$$\xi_i = \frac{\partial f(x)}{\partial x_i} \quad , \qquad i = 1,2,\ldots, n \quad ,$$

define the relationship between f and ϕ.

The function g is said to be closed in case the set $[g]$ is a closed set; for such a function, $\lim g(x) = -\infty$ when x approaches a boundary point of B which does not belong to B, and $\lim g(x) = f(y)$ when y is a boundary point belonging to B and x approaches y along a line segment interior to B. The function Mg is called the conjugate of g (without the restriction $\xi \geq 0$). The conjugate of a closed function is closed; also, conjugation is involutary, that is, $M(Mg) = g$. In the present context, the admissible functions f have envelopes $\bar{f}$ which are closed concave functions with domain $x \geq 0$. The maximum transform Mf (the conjugate of $\bar{f}$) is a closed concave (and bounded) function.

14.8 Optimal Distribution of Effort

In this section and the next, we consider two applications. The first is the problem of optimal distribution of effort. Let $x = (x_1, x_2, \ldots, x_n) \geq 0$ be a vector of n resources which is to be distributed among m activities as

$$x = x^{(1)} + x^{(2)} + \ldots + x^{(m)} \quad , \qquad x^{(j)} \geq 0 \quad ,$$

where

$$x^{(j)} = (x_1^{(j)}, x_2^{(j)}, \ldots, x_n^{(j)}) \ , \qquad j = 1,2,\ldots, m \quad .$$

Let the return from the j-th activity be $f_j(x^{(j)})$; then the problem is to solve

$$F(x) = \max_{\sum_j x^{(j)} = x} \sum_j f_j(x^{(j)}) \ , \qquad x^{(j)} \geq 0 \quad .$$

In the notation of maximum convolution this becomes

$$F = f_1 \oplus f_2 \oplus \ldots \oplus f_m \quad .$$

he binary operation $\oplus$ is associative and commutative. Applying M, e obtain

$$\Phi = \sum_{j=1}^{m} \phi_j \quad ;$$

hus,

(14.8.1) $$F = M\left(\sum_{j=1}^{m} \phi_j\right)$$

ives the maximum return in terms of the transforms $\phi_j = Mf_j$. Also $^{(j)}(y)$ with

(14.8.2) $$F^{(j)} = M\left(\sum_{k=1}^{j} \phi_k\right) \quad , \qquad F^{(m)} = F \quad ,$$

ives the maximum return from the first j activities when y is ssigned to this subset of activities. If the f_j are concave ncreasing, then (14.8.2) gives the true maximum return; otherwise, hey are the concave increasing caps of these return functions.

The determination of an optimal distribution is more ifficult; the following theorem is useful in this connection.

THEOREM 4. Let f, g be admissible concave increasing unctions; let $h = f \oplus g$. For a given value $x = \bar{x}$, let $u = \bar{u}$, $= \bar{v}$ be maximizing values of (14.5.1) so that

$$h(\bar{x}) = f(\bar{u}) + g(\bar{v}) \; , \qquad \bar{u} + \bar{v} = \bar{x} \quad .$$

hen every support direction of $[h]$ at $(\bar{x}, h(\bar{x}))$ is also a upport direction of $[f]$ at $(\bar{u}, f(\bar{u})$ and of $[g]$ at $(\bar{v}, g(\bar{v}))$.

PROOF. Let P, Q, R denote the points $(\bar{u}, f(\bar{u}))$, $(\bar{v}, g(\bar{v}))$, $\bar{x}, h(\bar{x}))$, respectively, where $P + Q = R$. Let t_R denote the angent cone to $[h]$ at P (this consists of all rays from P which are tangential* to the boundary of $[h]$). Now every ray of t_P lies below (i.e., below or on) t_R (considering $[f]$ translated to bring P into concidence with R). If not, there would be a tangential ray, to $[f]$ from P which lies strictly above t_R, and hence strictly above $[h]$; in that case there would be an increment $\delta\bar{u}$ such that $f(\bar{u} + \delta\bar{u}) + g(\bar{v}) > h(\bar{u} + \delta\bar{u} + \bar{v})$. This inequality is impossible since $h(x)$ maximizes $f(u) + g(v)$ for $u + v = x$. Next consider a hyperplane π through R with direction $(-\xi, 1)$; it is well known that π is a support of $[h]$ at R if and only if it is a support of t_R. Let π be such a support; it contains t_R below it. From the preceding, it follows that π translated to P contains t_P below it and is therefore a support of $[f]$. A similar argument applied to t_Q. This completes the proof.

This theorem suggests the following procedure for finding an optimal distribution $\bar{x}^{(j)}$ for a given initial $\bar{x}$. Determine a support direction $(-\xi, 1)$ to F at $(\bar{x}, F(\bar{x}))$. (This may be done without computing F by finding a point $(\bar{\xi}, \Phi(\bar{\xi}))$ on Φ with support direction $(-\bar{x}, 1)$.) Then for each j, find points of f_j where the support direction is also $(-\bar{\xi}, 1)$ (or, alternatively, find support directions $(-\bar{x}^{(j)}, 1)$ to ϕ_j at $\xi = \bar{\xi}$). This search is aided by (14.7.6); for example, if the x-space is two-dimensional, (14.7.6) reduces the search to examining the (one-dimensional) curve on $z = f_j(x_1, x_2)$ which satisfies

$$\bar{\xi}_1 x_1 + \bar{\xi}_2 x_2 - z = \phi_j(\bar{\xi}) \quad .$$

---

*Concavity implies derivatives in every direction from $\bar{x}$ when $\bar{x}$ is an interior point of $\{x | x_i \geq 0, \text{ all } i\}$; for $\bar{x}$ a boundary point, we are here assuming sufficient regularity that the functions have derivatives in every direction lying in the set.

14.9 A Multistage Allocation Process

As a second application, consider the following multistage process involving scalar variables. At the beginning of a time interval, an amount x (of some resource) is at hand; of this, any amount y, $0 \le y \le x$, may be committed so as to yield a return $g(y)$ (present value); the remainder $x - y$ is withheld for possible similar commitment at the beginning of the next period. At this latter point in time, however, we permit that the remainder has deteriorated (or been enhanced) by a factor, so that $x\, X(x - y)$ is actually available for allocation. The problem is to determine the maximum return over N (equal) time intervals.

Let $f_n(z)$ be the maximum return over n intervals beginning with an initial amount z, discounted to the initial time. Then the following functional equation of dynamic programming holds:

$$f_{n+1}(x) = \max_{0 \le y \le z} [g(y) + bf_n(a\cdot(x - y))] \quad , \quad f_1(x) = g(x) \quad ,$$

where b is the discount factor over a single time interval. Applying the maximum transform and using (14.7.5),

$$\phi_{n+1}(\xi) = \psi(\xi) + b\phi_n(\rho\xi) \ , \qquad \rho = \frac{1}{ab} \quad ,$$

where $\phi_n = Mf_n$, $\psi = Mg$. This has the explicit solution

$$\phi_N(\xi) = \psi(\xi) + b\psi\,(\rho\xi) + \ldots + b^{N-1}\,\psi(\rho^{N-1}\xi) \quad .$$

From this,

$$f_N(x) = (M^{-1}\,\phi_N)\,(x) = (M\phi_N)\,(x)$$

to within the concave increasing cap of f_N.

Optimal allocations $y_j = \bar{y}_j$ in this case of scalar variables for a given initial amount $x = \bar{x}_N$ may be determined recursively by solving

$$g(y_1) + bf_{N-1}(a\cdot(\bar{x}_N - y_1)) = f_N(\bar{x}_N) \quad ,$$

$$g(y_2) = bf_{N-2}(a\cdot(\bar{x}_{N-1} - y_2)) = f_{N-1}(\bar{x}_{N-1}) \, ,$$

$$\bar{x}_{N-1} = a\cdot(\bar{x}_N - \bar{y}_1) \quad ,$$

etc. Another method is to use Theorem 4 and (14.7.6). First determine $\xi = \bar{\xi}_N$ by solving

$$(\xi, \bar{x}_N) = \phi_N(\xi) + f_N(\bar{x}_N) \quad .$$

(To avoid computing f_N as the M-inverse of ϕ_N, one may find $\bar{\xi}_N$ as a place where ϕ_N has the support direction $(-\bar{x}_N, 1)$; this means a place where $d\phi_N(\xi)/d\xi = \bar{x}_N$, or, in case the derivative has a jump discontinuity, where the jump includes $\bar{x}_N$). Next, find $y_1 = \bar{y}_1$ such that

$$(\bar{\xi}_N, y_1) = \psi(\bar{\xi}_N) + g(y_1) \quad .$$

(Alternatively, solve $dg(y)/dy = \bar{\xi}_N$).

14.10 The General Transform T

We now leave the maximum transform M and its applications for the study of the general transform T satisfying (14.5.3). The principal goal is the proof of Theorem 5 and its corollary; the essential tool is Lemma 4 below. We return to the notation of Sec. 14.5.

LEMMA 3. The union of the convex hulls H_N, $N = 1,2,3,\ldots$ is H. Also,

$$\lim_{N\to\infty} \bar{f}_N(x) = \bar{f}(x) \; , \quad x \geq 0 \quad ,$$

where $\bar{f}_N$ is the cap of H_N. The sets H_N are closed.

PROOF. The convex hulls H_N of $[f_N^+]$ form an increasing sequence $H_N \subset H_{N+1}$ since the $[f_N^+]$ form such a sequence. It follows that the union, say K, of the H_N is a convex set; also, K contains each $[f_N^+]$, hence their union $[f^+]$, and hence the convex hull H of $[f^+]$. Conversely, H contains $[f^+]$, hence each $[f_N^+]$, hence each hull H_N, and hence their union K. Thus, $K = H$, as desired. To prove the second statement, notice that $\bar{f}_N(x)$ is monotone increasing with N and has a limit which is the upper bound of values z for which (x, z) belongs to the union of H_N, or H; but this is the definition of $\bar{f}(x)$. Finally, the last statement of the lemma is an immediate consequence of the fact that the convex hull of a bounded closed set is itself closed.

LEMMA 4. For any f,

$$f \oplus \bar{f} \oplus \ldots \oplus \bar{f} = \bar{f} \oplus \bar{f} \oplus \ldots \oplus \bar{f} \tag{14.10.1}$$

where $n + 1$ is the number of terms on each side.

PROOF. Part (i). First we show that it is sufficient to prove that

$$g \oplus \bar{g} \oplus \ldots \oplus \bar{g} = \bar{g} \oplus \bar{g} \oplus \ldots \oplus \bar{g} \text{ , where } g = f^+. \tag{14.10.2}$$

Since $\bar{g} = \bar{f}$, the right sides of (14.10.1) and (14.10.2) are the same; to show equality of the left sides it is enough to show that

$$f \oplus \bar{f} = g \oplus \bar{g} \quad .$$

The inequality $f \oplus \bar{f} \le g \oplus \bar{g}$ follows from $f \le g$. To show the reverse inequality, for a given $x \ge 0$ let

$$(g \oplus \bar{g})(x) = g(u) + \bar{g}(v)$$

$$= g(u) + \bar{f}(v) \text{ ,} \qquad u + v = x \quad .$$

By (14.6.9), for some $y \le u$, the right side equals

$$f(y) + \bar{f}(r) \leq f(y) + \bar{f}(v + u - y)$$

$$\leq (f \oplus \bar{f})(x) \quad .$$

This proves the equivalence of (14.10.1) and (14.10.2).

Part (ii). Since $y \leq \bar{g}$, the left side of (14.10.2) is $\leq$ the right side; hence, (14.10.1) will be established by showing that

$$(14.10.3) \qquad (g \oplus \bar{g} \oplus \ldots \oplus \bar{g})(x) \geq (\bar{g} \oplus \bar{g} \oplus \ldots \oplus \bar{g})(x) \quad .$$

The right side of (14.10.3) equals

$$\bar{g}(x^{(0)}) + \bar{g}(x^{(1)}) + \ldots + \bar{g}(x^{(n)}) \ , \quad \sum_{j=0}^{n} x^{(j)} = x \ .$$

All the $x^{(j)}$ may be taken as equal. For let $y = x/(n+1) = \Sigma_j a_j x^{(j)}$ with $a_j = 1/(n+1)$. Since $\bar{g}$ is concave.

$$\bar{g}(y) \geq (\tfrac{1}{n+1}) \ (\bar{g}(x^{(0)}) + \bar{g}(x^{(1)}) + \ldots + \bar{g}(x^{(n)})) \quad .$$

Thus,

$$(14.10.4) \qquad (\bar{g} \oplus \bar{g} \oplus \ldots \oplus \bar{g})(x) = (n + 1)\bar{g}(y)$$

where $(n + 1)y = x$.

Part (iii). Consider an arbitrary $\epsilon > 0$. By Lemma 3, $\lim \bar{g}_N(y) = \bar{g}(y)$ as $N \to \infty$. Pick an N such that

$$\bar{g}_N(y) > \bar{g}(y) - \epsilon/(n + 1)$$

with $0 \leq y_i < N$, $i = 1,2,\ldots, n$. Also, by Lemma 3, the convex hull H_x of $[g_N](=[f_N^+])$ is closed; therefore, the boundary point

$$(y, \bar{g}_N(y)) = (y, \bar{z})$$

belongs to H_N. It follows by a well known theorem on convex sets that there are points $(x^{(j)}, z^{(j)})$ in $[g_N]$ and weights $\alpha_j > 0$,

$\Sigma\alpha_j = 1$, such that

$$(14.10.5) \quad y = \sum_{j=0}^{k} \alpha_j x^{(j)} , \qquad \bar{z} = \sum_{j=0}^{k} \alpha_j x^{(j)} , \qquad 0 < k < n \quad .$$

We must have $z^{(j)} = g(x^{(j)})$ since otherwise, by pushing $z^{(j)}$ up to this value, a point of H_N could be obtained which lies above the boundary point $(y, \bar{z})$ of H_N.

Consider the case when $y_i > 0$, $i = 1,2,\ldots, n$. Since H_N is convex it has a supporting hyperplane H at $(y, \bar{z})$; further, this hyperplane is not vertical by the restriction $0 < y_i < N$. Consider the simplex determined by the k points $(x^{(j)}, z^{(j)})$; it has dimension $\leq n$ and is contained in H_N. The point of contact $(y, \bar{z})$ lies on H and is interior to the simplex (since $\alpha_j > 0$ for all j); therefore, the entire simplex lies on H. It follows that if (x, z) is any point of simplex, then $z = \bar{g}_N(x)$; that is

$$(14.10.6) \quad \bar{g}_N\left(\sum_{j=0}^{k} \beta_j x^{(j)} \right) = \sum_{j=0}^{k} \beta_j z^{(j)}$$

for all $\beta_j \geq 0$ with $\Sigma\beta_j = 1$. In particular, taking $\beta_j = 1$,

$$g_N(x^{(j)}) = z^{(j)} = \bar{g}_N(x^{(j)}) \quad .$$

Let the α_j of (14.10.5) be ordered so that α_0 is the largest. Then

$$\alpha_0 \geq 1/(k + 1) \quad .$$

Write

$$(k + 1)y = x^{(0)} + kx^* , \qquad x^* = \sum_{j=0}^{k} \beta_j x^{(j)}$$

where

$$\beta_0 = (1 + 1/k)\alpha_0 - 1/k \ , \qquad \beta_j = (1 + 1/k)\alpha_j \ ,$$

$$j = 1,2,\ldots, k$$

Then by (14.10.6) and (14.10.5)

$$g(x^{(0)}) + k\bar{g}_N(x^*) = z^{(0)} + k \, \Sigma \, \beta_j z^{(j)}$$

$$= (k + 1) \, \Sigma \, \alpha_j z^{(j)} \ ,$$

$$= (k + 1) \, \bar{g}_N(y) \quad .$$

Hence

$$g(x^{(0)}) + k\bar{g}_N(x^*) + (n - k) \, \bar{g}_N(y)$$

$$= (n + 1) \, \bar{g}_N(y)$$

$$> (n + 1) \, \bar{g}(y) - \varepsilon \quad ,$$

where $x^{(0)} + kx^* + (n - k) \, y = (n + 1) \, y = x$. Since $\bar{g} \geq \bar{g}_N$, when $\bar{g}_N$ is replaced by $\bar{g}$ in the first displayed line above, the resulting expression also exceeds the third displayed line. Accordingly,

$$(g \oplus \bar{g} \oplus \ldots \oplus \bar{g})(x) > (\bar{g} \oplus \bar{g} \oplus \ldots \oplus \bar{g})(x) - \varepsilon \quad .$$

Since ε is arbitrary, the desired inequality (14.10.3) follows.

This establishes (14.10.3) for all x with $x_i > 0$, $i = 1,2,\ldots, n$. Since both sides of (14.10.3) are continuous functions of x, the same holds for any $x \geq 0$. This completes the proof.

THEOREM 5. Let T carry each admissible function f into some real-valued transform function $\phi = Tf$. Suppose that for any admissible functions f, g

$$T(f \oplus g) = Tf + Tg$$

in the sense that the transform on the left has as its domain of definition the common domain of the transform on the right and the equality holds there. Then for any admissible f and its concave increasing envelope $\bar{f}$, the domain of definition of Tf contains that of $T\bar{f}$ and the equality

$$Tf = T\bar{f}$$

holds on the domain of $T\bar{f}$.

PROOF. Let D and E be the domains of definition of Tf and $T\bar{f}$, respectively. Consider Eq. (14.10.1) of Lemma 4. From the assumption of the theorem the transform of the left side is

$$Tf + T\bar{f} + \ldots + T\bar{f}$$

and has as its domain of definition the intersection of D and E; also the transform of the right side is

$$T\bar{f} + T\bar{f} + \ldots + T\bar{f}$$

and has domain E. Since the two sides are equal, their transformations are equal and have the same domain of definition. Hence D contains E and the conclusion of the theorem follows.

COROLLARY. Let T satisfy the hypotheses of Theorem 5. Then T factors into

$$T = \lambda M$$

where λ is an operator on the space of image functions of M which satisfies

$$\lambda(\phi_1 + \phi_2) = \lambda\phi_1 + \lambda\phi_2 \quad .$$

PROOF. We have $T = TMM$; hence $T = \lambda M$ with $\lambda = TM$. Let $\phi_1 = Mf_1$, $\phi_2 = Mf_2$. Then $\phi_1 + \phi_2 = M(f_1 \oplus f_2)$. Hence

$$TM(\phi_1 + \phi_2) = TMM(f_1 \oplus f_2) = T(\bar{f}_1 \oplus \bar{f}_2)$$

$$= T\bar{f}_1 + T\bar{f}_2 = TM\phi_1 + TM\phi_2 \quad ,$$

which proves the theorem.

14.11 Minimum and Multiplicative Convolution

We return to the maximum transformation M and summarize the necessary modifications for the related convolutions

(14.11.1) $h(x) = \min_{u+v=x} \{f(u) + g(v)\}$,

(14.11.2) $H(x) = (f \oplus G)(x) = \max \{F(u) \times G(v)\}$

and (14.11.2) with "max" replaced by "min", we assume $F > 0$, $G > 0$.

The first may be cast into a maximum convolution by replacing the functions by their negative; (14.11.1) is equivalent to

$$-h(x) = \max_{u+v=x} \{-f(u) - g(v)\}$$

$$= ((-f) \oplus (-g))(x) \quad .$$

Admissible functions in this context are ones whose negatives are admissible in the earlier sense. We specify the minimum convolution Nf as $-M(-f)$; that is,

$$(Nf)(\xi) = -\inf_{x \geq 0} \{(\xi, x) + f(x)\} \quad ;$$

this is defined for all directions $(-\xi, -1)$, $\xi \geq 0$, in which $f(x)$ is bounded. Then

(14.11.3) $Nh = Nf + Ng.$

For problem (14.11.1), $[f]$ is taken as the set of (x, z) such that $z \geq f(x)$. With f is associated its (lower) convex decreasing envelope $\underline{f}$; we have

(14.11.4) $\quad Nf = N\underline{f}$.

Also, Nf is a convex decreasing function, and

$$N(Nf) = N\underline{f} \quad .$$

Property (14.11.4) holds for any transformation that satisfies (14.11.3)

The multiplicative problem (14.11.2) is converted to an additive one by introducing f, g, h where

$$F(x) = e^{f(x)}, \quad G(x) = e^{g(x)}, \quad H(x) = e^{h(x)} .$$

Then (14.11.2) is equivalent to

$$h(x) = (f \oplus g)(x) \quad .$$

Introduce the transformation

$$\mathcal{M}F = e^{-Mf} \quad .$$

Explicitly,

$$(\mathcal{M}F)(\xi) = \sup_{x \geq 0} \{e^{-(\xi,x)}F(x)\} \quad .$$

Since $Mh = Mf + Mg$, we have the property

(14.11.5) $\quad \mathcal{M}(F \oplus G) = (\mathcal{M}F) \times (\mathcal{M}G)$.

(Notice that the earlier restriction on the growth of an admissible function is now that $F(x)$ be dominated by e^{kr} for arbitrary $k > 0$, where r is the length of x). We have $\mathcal{M}e^{f} = \mathcal{M}e^{\bar{f}}$; this property holds for any transformation satisfying (14.11.3)

To invert the transformation we use

$$\eta S = e^{Ns} \quad \text{where} \quad S(\xi) = e^{s(\xi)} ,$$

that is,

$$(\eta S)(x) = \inf_\xi \{e^{(x,\xi)} S(\xi)\} \quad .$$

Then

$$\eta(\mathcal{M}F) = e^{-N(-Mf)} = e^{MMg} = e^{\bar{f}} \quad .$$

For the multiplicative case (14.11.2) with "min" the transformation η replaces $\mathcal{M}$, and $\mathcal{M}$ inverts η in the sense that $\mathcal{M}(\eta F) = e^{\underline{f}}$.

It may be remarked that in (14.11.1) the "maximum transform" for the original convolution $\oplus$ was chosen as $-Mf$. That is,

$$\mu(\xi) = \sup_{x \geq 0} \; [-(\xi, x) + f(x)] \quad ;$$

this makes the transform convex, but serves as well as M for applications; it is inverted by

$$\bar{f}(x) = \inf_\xi [(x, \xi) + \mu(\xi)] \quad .$$

14.12 Examples of Transforms

We conclude with some explicit examples of maximum transforms of functions of a single variable. The first list (14.12.1) to (14.12.5) gives pairs $f, \varphi = Mf$; the second list (14.12.6) to (14.12.9) gives pairs $F, \Phi = \mathcal{M}F$ in which correspondingly numbered formulas are obtained by

$$F = e^f \; , \qquad \Phi = e^{-\varphi} \quad .$$

It is understood that $b \geq 0$.

(14.12.1)

$$f(x) = \begin{cases} a + bx & \text{for } \; 0 \leq x \leq x_0 \quad , \\ a + bx_0 & \text{otherwise,} \end{cases}$$

$$\phi(\xi) = \begin{cases} -a + x_0 \, (\xi - b) & \text{for } \; 0 < \xi < b \quad , \\ -a & \text{otherwise.} \end{cases}$$

$$f(x) = a + b\log x \quad ,$$

(14.12.2)

$$\phi(\xi) = -a + b(1 + \log\frac{\xi}{b}) \quad .$$

$$f(x) = \begin{cases} a + b\log x & \text{for } 0 \le x \le x_0 \quad , \\ a + b\log x_0 & \text{otherwise} \end{cases}$$

(14.12.3)

$$\phi(\xi) = \begin{cases} -a + b\left[\frac{\xi}{\xi_0} + \log\frac{\xi_0}{b}\right] & \text{for } 0 \le \xi \le \xi_0 \quad , \\ -a + b\left[1 + \log\frac{\xi}{b}\right] & \text{otherwise} \quad , \end{cases}$$

where $\xi_0 = b/x_0$.

$$f(x) = bx^{1/p} \quad , \qquad 0 < 1/p < 1 \quad ,$$

$$\phi(\xi) = -\frac{b}{p}\left(\frac{b}{p\xi}\right)^{q-1} \quad , \quad \text{where } 1/p + 1/q = 1 \quad .$$

(14.12.4)

$$f(x) = \begin{matrix} bx^{1/p} & \text{for } 0 \le x \le x_0 \quad , \\ bx_0^{1/p} & \text{otherwise,} \end{matrix}$$

(14.12.5)

$$\phi(\xi) = \begin{cases} b\left(\frac{1}{p} \cdot \frac{\xi}{\xi_0} - 1\right)\left(\frac{b}{p\xi_0}\right)^{q-1} & \text{for } 0 \le \xi \le \xi_0 \quad , \\ -\frac{b}{q}\left(\frac{b}{p\xi}\right)^{q-1} & \text{otherwise} \quad , \end{cases}$$

where $\xi_0 = \frac{b}{px_0^{1/q}}$, p, q as in (14.12.4).

$$F(x) = \begin{cases} Ae^{bx} & \text{for } 0 \le x \le x_0 \quad , \\ Ae^{bx_0} & \text{otherwise,} \end{cases}$$

(14.12.6)

$$\Phi(\xi) = \begin{cases} Ae^{x_0(b-\xi)} & \text{for } 0 \le \xi \le b \quad , \\ A & \text{otherwise.} \end{cases}$$

(14.12.7) $$F(x) = Ax^b \quad ,$$

$$\Phi(\xi) = Ae^{-b}\left(\frac{b}{\xi}\right)^b \quad .$$

$$F(x) = \begin{cases} Ax^b & \text{for } 0 \le x \le x_0 \quad , \\ Ax_0^b & \text{otherwise,} \end{cases}$$

(14.12.8)

$$\Phi(\xi) = \begin{cases} Ae^{-b\xi/\xi_0}\left(\frac{b}{\xi_0}\right)^b & \text{for } 0 \le \xi \le \xi_0 \quad , \\ Ae^{-b}\left(\frac{b}{\xi}\right)^b & \text{otherwise.} \end{cases}$$

(14.12.9) $$F(x) = e^{bx^{1/p}} \quad ,$$

$$\Phi(\xi) = \exp\left[\frac{b}{q}\left(\frac{b}{p\xi}\right)^{q-1}\right] \quad .$$

Similarly, (14.12.10) has an analog obtained by exponentiation of the functions displayed in (14.12.10). An interesting special case of (14.12.9) occurs when $p = 2$; then

(14.12.10) $$F(x) = e^{b\sqrt{x}} \quad , \qquad \Phi(\xi) = e^{b^2/4\xi} \quad .$$

14.13 The Maximum Transform and Semigroups of Transformations

The problem of determining the maximum of the function

$$(14.13.1) \qquad F(x_1, x_2, \ldots, x_N) = \sum_{i=1}^{N} g_i(x_i)$$

over the domain D_N defined by $\sum_{i=1}^{N} x_i = x$, $x_i \geq 0$, is one with various ramifications and applications. Analytic solutions and computational algorithms have been obtained in a number of ways. Let us now discuss a new way of generating solutions of (14.13.1). Let $g(x, a)$ be a scalar function of the scalar variable x and the M-dimensional vector a with the group property that

$$(14.13.2) \qquad \max_{x_1 + x_2 = x} [g(x_1, a) + g(x_2, b)] = g(x, h(a, b))$$

$$(x_1, x_2 \geq 0) \quad ,$$

where $h(a, b)$ is a known function of a and b. It follows inductively that

$$(14.13.3) \qquad \max_{D_N} \left[\sum_{k=1}^{N} g(x_k, a^{(k)}) \right] = g(x, h(a^{(1)}, a^{(2)}, \ldots, a^{(N)})) \quad ,$$

where D_N is as above, and $h(a^{(1)}, a^{(2)}, \ldots, a^{(N)})$ is obtained from $h(a, b)$ in a recurrent fashion. The function $g(x, a) = ax^p$, $0 < p \leq 1$, with $a \geq 0$ is a function of the desired type. How can we generate classes of functions with this property, and can we determine all of them?

In previous sections, we discussed the transform

$$(14.13.4) \qquad F(y) = M(f) = \max_{x \geq 0} [f(x) - xy]$$

a transform which plays a basic role in the study of convexity. This transform possesses the important dissolving property

$$(14.13.5) \qquad M[\max_{x_1 + x_2 = x} [g(x_1, a) + g(x_2, b)]] = M(g(x, a) + M(g(x, b)) \quad .$$

Taking advantage of this relation, we can obtain functions satisfying (14.13.2) by starting with functions $G(x, a)$ satisfying the simpler relation

$$(14.13.6) \quad G(x, a) + G(x, b) = G(x, h(a, b)) ,$$

and inverting

$$(14.13.7) \quad g(x, a) = M^{-1}(G(x, a)) = \min_{y \geq 0} [G(y, a) + xy] .$$

14.14 Solutions of the functional equation

If a is an M-dimensional vector with components $a_1, a_2, \ldots, a_M$, a very simple class of solutions (14.13.6) is given by

$$(14.14.1) \quad G(x, a) = (a, G(x)) = \sum_{i=1}^{M} a_i G_i(x) ,$$

$$h(a, b) = a + b .$$

Under various assumptions of analyticity, it may be shown that aside from inessential changes of variable, $a \rightarrow \phi(a)$, these are the only solutions.

14.15 Parametric Representation

If it is possible to obtain the minimum value in (14.13.7) by means of differentiation, we obtain the parametric representation

$$(14.15.1) \quad g(x, a) = G(y, a) + xy ,$$

$$x = -G_y(y, a) .$$

Assuming that $G(x, a)$ is given by (14.14.1), we face the interesting problem of determining the a_i and $G_i(x)$ so as to fit a given function $g(x)$ in some optimal fashion. Having done this, we can find quick and useful solutions to the original variational problem,

(14.13.1) *et seq*. The point is that in this way we find exact solutions to approximate problems as opposed to the usual approximate solution to an exact problem.

Bibliography and Comments

Section 14.1. Allocation processes are treated in Bellman, R., and Dreyfus, S., Applied Dynamic Programming, Princeton University Press, Princeton, NJ, 1962.

In this book, the computational approach of dynamic programming is given in detail. In connection with the Hitchcock-Koopmans transportation problem, we show how successive approximations may be used to treat allocation processes.

Section 14.2. These results were given in Bellman, R., "Dynamic Programming and Lagrange Multipliers", Proceedings of the National Academy of Sciences, Vol. 42, No. 10, October 1956, pp. 767-769.

Section 14.5. We are following the paper, Bellman, R., and Karush, W., "Mathematical Programming and the Maximum Transform", J. Soc. Indust. Appl. Math., Vol. 10, No. 3, 1962, pp. 550-567.

Other applications of the maximum transform may be found in Bellman, R., and Karush, W., "On the Maximum Transform and Semigroups of Transformations", Bulletin American Mathematical Society, Vol. 68, 1962, pp. 516-518.

Bellman, R., and Karush, W., "On the Maximum Transform", Journal of Mathematical Analysis and Applications, Vol. 6, 1963, pp. 67-74.

The maximum transform has the same relation to the conjugate transform of Fenchel as the Laplace transform has to the Fourier transform.

See also Rockafellar, T., Convex Analysis, Princeton University Press, Princeton, NJ, 1970.

Section 14.13. These results were given in Bellman, R., and Karush, W., "On the Maximum Transform and Semigroups of Transformations", Bulletin of the American Mathematical Society, Vol. 68, No. 5, Sept. 1962, pp. 516-518.

APPENDIX A

A Function is a Mapping — Plus a Class of Algorithms

14.A.1 Introduction

One of the fundamental concepts of mathematics is that of function. The standard definition involves the notion of mapping of correspondence. Given two sets S_1 and S_2, a correspondence which associates each element of S_1 with a subset of elements of S_2 defines a function whose domain is S_1.

The purpose of this note is to point out that, important as this definition is, it is incomplete. In order to make it complete, we must add, or adjoin, all possible algorithms for the realization of the mapping. Function, like number, is a class concept. This question of realizability is, of course, a classic one, first emphasized by Kronecker and discussed extensively at the turn of the century.

At that time, however, the emphasis was on the conceptual and philosophical aspects. With the development of the digital computer, realizability translates into algorithmic feasibility. As soon as we accept the fact that any mapping can be realized in a number of different ways, we automatically create a number of novel, interesting, and significant mathematical problems. Basically, the question is that of devising particular algorithms for particular purposes.

To illustrate what we mean without attempting to make the notion rigorous, let us discuss three problems in analysis. The first is classic, considered by Cramer and then Gauss; the second was first treated by Horner and recently resolved by Pan [1]; the third, apparently the simplest, is still open.

14.A.2 Linear Systems of Algebraic Equations

Consider the linear system

$$\text{(A.2.1)} \qquad \sum_{j=1}^{N} a_{ij}x_j = b_i \ , \qquad i = 1,2,\ldots, N \quad ,$$

under the assumption that the matrix of coefficients is non-singular. In this case, the equation defines N linear functions of the b_i, in vector form

$$\text{(A.2.2)} \qquad x = A^{-1}b \quad .$$

One immediate way of realizing these functions is by use of determinants and Cramer's rule. This is not a feasible approach for large N. Gauss pointed out that a straightforward elimination technique was feasible since it involves of the order of N^3 multiplications as opposed to order of $N!$ multiplications for Cramer's rule. We focus on multiplications since they require so much more time than additions. More recently, the problem of a suitable order of elimination of variables has been considered, the "pivoting" technique.

The variables are eliminated in a sequence that is determined by the relative values of certain minors of $|A|$. This is a particular case of the general problem of utilizing characteristics of the matrix A to calculate the solution of (A.2.1). Related to this is the question of determining efficient algorithms for calculating only a few of the x_i. Monte Carlo techniques can be useful here.

What we wish to inducate without going into details is that this apparently simplest of equations possesses an unlimited number of associated problems as soon as we ask the appropriate questions.

14.A.3 Evaluation of a Polynomial

Consider the polynomial

$$\text{(A.3.1)} \qquad p_N(x) = a_0x^N + a_1x^{N-1} + \ldots + a_N \quad .$$

It can be evaluated for a particular value of x by means of $2N-1$ multiplications. Can we evaluate it using a smaller number of multiplications? Horner suggested the algorithm

$$p_N(x) = ((a_0x + a_1)\, x + a_2)x + \dots , \tag{A.3.2}$$

which requires only N multiplications. For many years, the problem was open as to whether this was a best possible procedure as far as the number of multiplications was concerned. Ostrowski established this for polynomial of degree four or less. Recently, Pan [1] proved that this was a best possible estimate for a general p.

This result by no means ends the investigation. First of all, there is the problem of determining an algorithm that might be used to evaluate efficiently a number of different functional values. Clearly, it might be worthwhile to spend some effort constructing some auxiliary functions if one were to carry out a number of evaluations. Second, there is the problem of constructing an algorithm for specific polynomials of special form. Third, there is the problem of evaluating polynomials of two variables, etc.

14.A.4 Evaluation of an Exponential

The exponential 2^n can be evaluated using $(n-1)$ multiplications. However, it is clear that we can do very much better by combining powers of the form $2, 2^2, 2^4,$ and so on. What is the most efficient way if we measure efficiency as before in terms of the number of multiplications needed? Remarkably, the problem is unsolved. What is more, we do not even possess a good search algorithm for solving the problem for any particular value of n.

14.A.5 Discussion

The three examples above illustrate how many different types of problems arise in an easy fashion if we attempt to grade the algorithms that realize a function in terms of a preassigned criterion. Sometimes, as above, this criterion is time, time required to carry out multiplications. Sometimes, the criterion is accuracy. Most frequently in connection with digital computers, the criterion is rapid-access storage capacity, and then time.

Vast classes of problems of the foregoing nature that arise in connection with differential equations of ordinary and partial type, as well as functional equations of more complex type, have been investigated using only a very restricted class of algorithms, e.g., those based on conventional difference equations. Furthermore, as usual in analysis, the most interesting problems are not those of exact solution but rather of approximate solution. How close can we come to an algorithm of a specified nature by means of algorithms of other types? A great deal of work has been done in connection with functional approximation, but very little in connection with algorithmic approximation. A particular example of this kind of question is furnished by the Mascheroni construction, use of compass alone or ruler alone, rather than free use of both, [2]. All of this is preliminary to a theory of closure of operations, and all of it stems from the extended concept of function proposed above.

References

1. V. Ja. Pan, On Means of Calculating Values of Polynomials, Uspechi Mat. Nauk 21, No. 1(127), 1966, pp. 103-134.

2. Bellman, R., On Some Mathematical Recreations, American Math. Monthly 69, 1962, pp. 640-643.

APPENDIX B

On Proving Theorems in Plane Geometry via Digital Computers

14.B.1 Introduction

The development of the digital computer has focused considerable attention upon various types of algorithms, and, in particular, upon those connected with logical processes and decision-making. An offshoot of this has been the set of attempts by various people, with varying degrees of success, to replicate human thought processes with the aid of a digital computer. In this connection, let us cite the work in translation of languages, pattern recognition, chess playing, checker playing, and the proving of logical and geometric theorems.

In pursuing these goals, there are many different approaches that can be pursued. At one extreme, we can imitate what the human mind does; at the other extreme, we can fasten our attention solely upon the capabilities of an analog or digital computer. In between, we have a continuum of man-machine processes. Since it is generally agreed by knowledgeable people that we possess very little understanding of the working of the brain, it is clearly hazardous to follow the first route. We shall restrain ourselves exclusively to a completely rigorous use of the computer in establishing geometric theorems.

The basic idea is quite simple. The structure of Euclidean plane geometry permits us to express geometric theorems as algebraic identities. Algebraic identities can be established by verification of a sufficiently large number of cases. This verification should be possible, purely arithmetically, using a digital computer. There are, however, many interesting problems associated with a procedure of this type, as we shall see below.

Let us hasten to add that we see no intrinsic value to establishing geometric theorems in this fashion. We do feel that it is pedagogically of some value to have the student interested in the uses of the computer and to try his skill at problems of this type, and there are some nontrivial associated arithmetic and analytic questions. Furthermore, numerous questions arise as to what we mean by "fundamental theorems" of geometry. All of this will be illustrated by the following discussion.

14.B.2 The Medians of an Isosceles Triangle

Suppose that we wish to prove that the medians to the equal sides of an isosceles triangle are equal. Considering the figure below, we must establish the identity

$$\text{(B.2.1)} \quad \left[\left(a - \left(-\frac{a}{2}\right)\right)^2 + \left(0 - \frac{b}{2}\right)^2\right]^{1/2} = \left[\left(\frac{a}{2} - (-a)\right)^2 + \left(\frac{b}{2} - 0\right)^2\right]^{1/2}$$

This is, of course, obvious upon inspection; but with a digital computer, we are not allowed "inspection".

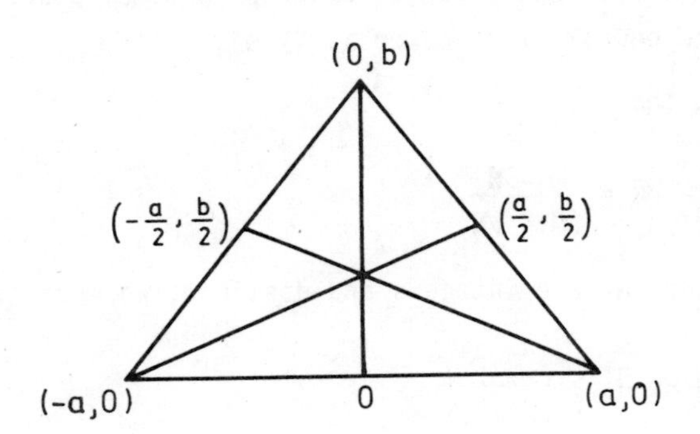

As far as the computer is concerned, an isosceles triangle is determined by the three vertices (0, b), (a, 0), (-a, 0), and the lengths of the medians are determined by simple algorithms which provide first the midpoints and then the length according to the distance formula. We are assuming the Pythagorean theorem which enables us to use the usual distance formula of analytic geometry.

We establish the equality of the two sides of (B.2.1) in a two-step process. We first invoke some general algebraic theorems which assure us that it is sufficient to verify equality for a finite set of values of a and b, and then use the computer to carry out this arithmetic confrontation.

There are several ways to proceed. Let us sketch one. Since the expressions are homogeneous in a and b (scale is unimportant), it suffices to take $b=1$. Since equaltiy of the squares implies equality of positive quantities, let us square both sides. It remains to establish the identity of two quadratic polynomials in a. For this, equality at <u>three</u> values of a suffices. Choose three convenient values, e.g., $a = 2, 4, 6$.

At this point, the reader may justifiably worry about round-off error. After all, computer arithmetic is not ordinary arithmetic. Suppose that we had not thought of the artifice of squaring, or, in general, of rationalizing. How would we establish that $M=N$ by comparing the calculated values of $\sqrt{M}$ and $\sqrt{N}$? Does agreement of M and N to a sufficiently large number of decimal places assure us that they are equal? The answer is "yes".

Observe that

$$\text{(B.2.2)} \qquad \sqrt{M} - \sqrt{N} = \frac{M - N}{\sqrt{M} + \sqrt{N}} \quad .$$

Hence, if M and N are integers and distinct, we must have

$$\text{(B.2.3)} \qquad |\sqrt{M} - \sqrt{N}| \geq \frac{1}{\sqrt{M} + \sqrt{N}} \quad .$$

If arithmetic calculations show that

$$\text{(B.2.4)} \qquad |\sqrt{M} - \sqrt{N}| \leq \varepsilon < \frac{1}{\sqrt{M} + \sqrt{N}} \quad .$$

we can conclude that $M=N$. Starting with the values of M and N, we know how to obtain the accuracy of (B.2.4).

In the general case, a number of interesting questions arise as to the number of verifications required and the methods to be used to obtain this verification. How, for example, does one systematically reduce a problem involving distances to a polynomial identity?

14.B.3 Discussion

We leave it to the reader to investigate the possibility of establishing the existence of the Euler line, the Simpson line, the nine-point circle, and so on. It is also clear that we can "generate" theorems of this type in a completely uninspired tabulating sets of points and lines and testing collinearity, coincidence, etc. Some of this would be pursued in an adaptive fashion, as, for example, the search for the nine-point circle. As mentioned previously, none of this has any intrinsic interest. But these are useful problems for training purposes, since they deal with familiar types of questions requiring no advanced training. They illustrate what is meant by the term "algorithm", and they also demonstrate the value of ingenuity and knowledge of structure of a process.

SUBJECT INDEX